1·2·3 Math

Pre-Math Opportunities for Working With Young Children

Written by Jean Warren

Illustrated by Marion Hopping Ekberg

Totline
PUBLICATIONS

Warren Publishing House
A Division of Frank Schaffer Publications
Torrance, California

Some of the activity ideas in this book were originally submitted to the Totline Newsletter by other teachers. We wish to acknowledge the following contributors: Janice Bodenstedt, Jackson, MI; Vicki Claybrook, Kennewick, WA; Paula Foreman, Lancaster, PA; Sue Foster, Mukilteo, WA; Judy Hall, Wytheville, VA; Mildred Hoffman, Tacoma, WA; Colraine Pettipaw Hunley, Doyelstown, PA; Karel Killimnik, Philadelphia, PA; Barb Mazzochi, Villa Park, IL; Joleen Meier, Marietta, GA; Kathy Monahan, Breckenridge, MN; Susan M. Paprocki, Northbrook, IL; Susan Peters, Upland, CA; Jeanne Petty, Camden, DE; Lois E. Putnam, Pilot Mountain, NC; Nancy Ridgeway, Bradford, PA; Betty Silkunas, Lansdale, PA; Jane M. Spannbauer, So. St. Paul, MN; Kristine Wagoner, Puyallup, WA; Nancy C. Windes, Denver, CO; Saundra Winnett, Fort Worth, TX.

Editorial Staff:
Gayle Bittinger, Kathleen Cubley, Brenda Lalonde,
Elizabeth McKinnon, Erica West

Art and Production Staff:
Production Manager: Jo Anna Brock
Art Manager: Jill Lustig
Design: Kathy Kotomaimoce
Computer Graphics: Sarah Ness
Cover Design: Eric Stovall

ISBN 0-911019-52-9

Library of Congress Catalog Number 92-80528
Printed in the United States of America
Published by: Warren Publishing House
Editorial Office: P.O. Box 2250
Everett, WA 98203
Business Office: 23740 Hawthorne Blvd.
Torrance, CA 90505

20 19 18 17 16 15 14 13 12 11 10 9 8 7 6

Introduction

1•2•3 Math is basically divided into two parts. The first part provides a step-by-step introduction to basic pre-math skills. The second part centers around opportunities for incorporating pre-math activities in play situations.

As with all the books in Totline's 1•2•3 Series, *1•2•3 Math* is filled with open-ended, no-lose activities. It offers opportunities for children to acquire pre-math skills through hands-on situations that are both meaningful and fun.

Young children need everyday opportunities to develop math concepts. You can facilitate this need by helping your children discover math relationships all around them. From sorting to matching to measuring, from time to phone numbers to ages, all offer practical situations for young children to experience number concepts.

Remember to keep all experiences relaxed, fun and non-threatening. Encourage problem-solving by stepping back occasionally and letting your children discover number concepts on their own.

As you work with the units in this book, feel free to adapt or add activities as desired. Also, make a point of vocalizing number concepts for your children in ways such as these: "Fill the glasses half full." "We will eat lunch at 12 o'clock." "Joseph is three. Katie is four. Katie is older than Joseph." "Would you bring me a block from the tallest shelf?" "Put the crayons back into the middle box."

Opportunities to teach pre-math skills are all around us. We as teachers and parents need only be aware of these opportunities and incorporate them naturally into the lives of our children.

Jean Warren

Contents

Sorting

Color Boxes

For a simple beginning sorting activity, have your children sort objects by color. Start by choosing two colors such as red and blue. Help the children collect small red and blue objects from around the room. Set out two boxes, one colored red and one colored blue. Then let the children sort the objects that they collected into the matching colored boxes. As your children become familiar with the activity, substitute or add different colored boxes.

Sorting Fun

Collect several different kinds of small objects in three or more different colors (crayons, toy cars, beads, blocks, etc.). Place the objects in a box and mix them up. Sit with one or two children at a time. Ask them to take objects out of the box and sort them into different groups by kind. Then put the objects back into the box, mix them up again and ask the children to sort them into groups by color. For an additional sorting activity, have the children group the objects by size.

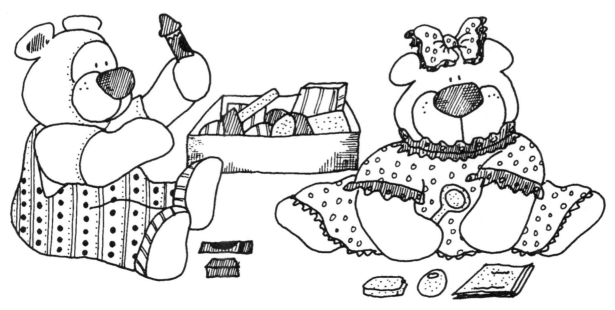

Animal Sort

Select a large shoebox with a lid. Divide the inside of the lid into three sections. Using felt-tip markers, draw a land scene in the first section, a sky scene in the second section and a water scene in the third section. Draw or glue pictures of different kinds of animals on small cards. Talk with your children about the scenes drawn on the shoebox lid. Then have them sort the animal cards into three groups: animals that walk, animals that fly and animals that swim. When the children have finished playing, ask them to put the cards back into the shoebox.

Variation: Adapt the game so that your children can sort the cards into these categories: farm animals, wild animals and pet animals.

Button Box

Select a small shallow box. Inside the box, divide the bottom into four sections and glue a piece of different colored construction paper in each section. Set out buttons that are the same four colors. Let your children take turns placing the buttons in the matching colored sections in the box.

Variation: Use a muffin tin instead of a box and place different colored circles of construction paper in the bottoms of the cups.

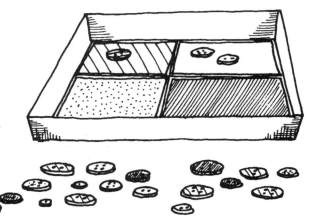

Shape Sorting Box

For this activity you will need a sturdy box that has a removable lid. Using a craft knife, cut a circle, a square and a triangle out of the lid. Cut a number of matching shapes out of sturdy cardboard. Place the lid on the box and give the cut-out shapes to your children. Let them take turns putting the shapes into the box through the matching shaped holes.

Variation: Cut squares and circles out of the box lid that are the size of small wooden blocks and empty thread spools. Have your children put the blocks and spools into the box through the appropriate holes.

Sorting Playing Cards

Remove the face cards from a deck of playing cards. Then use the deck to play a sorting game. Sit on the floor with several children at a time. Place one red card and one black card face up on the floor and put the remainder of the deck face down. Let the children take turns drawing cards from the top of the deck and sorting them by color into two piles. Follow the same procedure to help the children sort the cards by suit and by number.

Categorizing Toys

Use this activity to help keep your toy area tidier. After playtime one day, ask your children to divide their toys into two groups: small toys and large toys. Have the children put the small toys into a box or drawer and the large toys on a shelf or against a wall. On later days, ask the children to divide their toys into other categories such as hard toys and soft toys or wooden toys and plastic toys. After you have done this activity a few times, let the children help decide which way of grouping their toys works best. Have them use that arrangement when they put their toys away each day.

Sorting Shoes

Collect 10 to 20 pairs of shoes. Include a wide variety of kinds and sizes. Place the shoes in a large box. Each day empty the box onto the floor and let your children sort the shoes in a different way. For example, start by having them sort the shoes into pairs. Then let them continue sorting the shoes by color, size, texture, type of fastener, etc.

Seasonal Sort

Divide a piece of butcher paper into four sections. Label the sections with pictures to indicate the four seasons: a flower for spring, a sun for summer, a pumpkin for fall and a snowman for winter. On index cards, glue or draw pictures of spring scenes, summer scenes, fall scenes and winter scenes. Mix up the picture cards and let your children take turns placing them in the appropriate sections on the butcher paper.

Variation: Label the sections on the butcher paper to represent rooms in a house. Let your children sort pictures of things found inside a house into the appropriate rooms.

Box Sort

Collect an assortment of small boxes. Place the boxes on a table or on the floor. Talk about different ways that the boxes might be sorted. Then help your children group the boxes according to such things as color, size, shape, construction and former contents.

Transportation Sort

Cut out pictures of vehicles that travel on land, vehicles that travel on water and vehicles that travel in air. Mount the pictures on heavy paper and mix them up. Let your children sort the pictures into three piles according to where the vehicles travel.

Variation: Draw a land scene, a water scene and a sky scene on a piece of butcher paper. Place the paper on a table or on the floor. Let your children sort toy vehicles into groups by placing them on the appropriate scenes on the butcher paper.

Clothesline Sorting Game

Tie a clothesline between two chairs and set out a box of clothespins. Fill a basket with a variety of fabric squares. Let your children sort the squares by clipping them to the clothesline. For example, have them hang up all the squares that contain the color green, all the flowered squares or all the yellow squares.

Variation: Fill the basket with clothing shapes cut from fabric. Ask your children to hang up all the "stockings," all the "shirts," etc.

Counting

Counting Cotton Balls

Let your children line up cotton balls and count them. (Or use other small objects such as peanuts, buttons or straw sections.) Then let the children play one or more of the following counting games.

- Place a number of cotton balls in a paper bag. Let your children each have a turn reaching into the bag and grabbing a handful. Then count together how many cotton balls each child has taken.

- Let your children fill yogurt cups, plastic sandwich bags or small boxes with cotton balls. Then have them empty their containers and count the number of cotton balls that were inside.

- Use large index cards to make a set of five counting cards. Draw one circle on the first card, two circles on the second card, and so on. Lay the cards on a table and set out 15 cotton balls. Let your children take turns counting the cotton balls as they place them in the circles on the cards.

Stringing and Counting

Whenever your children are stringing objects such as beads, macaroni or straw sections, turn the activity into a counting experience. For example, give each child a shoelace (knotted at one end) and eight to ten buttons with large holes. Have the children count their buttons before stringing them on their shoelaces. Then have them count their buttons while they are stringing them or when they have finished stringing them. If desired, encourage your children to count their buttons as they unstring them also.

Felt Leaf Game

Cut a bare tree shape out of brown felt and place it on a flannelboard. Cut out 10 to 20 small felt leaf shapes. Ask a child to place a particular number of leaves on the tree. Then with the group, count the number of leaves as you remove them. Ask the next child to place a different number of leaves on the tree. Continue until each child has had a turn.

Variation: Find a long brown glove (or use a felt-tip marker to color a long white glove brown). Attach loops of masking tape, sticky sides out, to the backs of the felt leaf shapes. Slip on the glove to turn your hand and arm into a "tree" and let your children attach the leaves to the finger "branches."

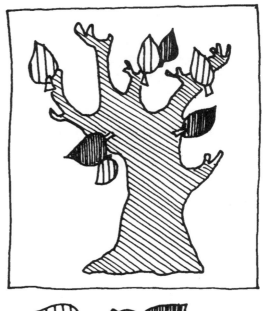

Counting Rhymes and Songs

Cut seasonal shapes (pumpkins, hearts, bunnies, etc.) out of felt. Place the shapes on a flannelboard. Let your children manipulate the shapes as you recite simple counting rhymes like the one below.

Here is a pumpkin,
And here is a pumpkin,
And another pumpkin I see.
Shall we count them?
Are you ready?
One, two, three!

Traditional

Extension: As your children become more proficient in counting, encourage them to sing and count as follows:

Sung to: "Ten Little Indians"

One little, two little, three little pumpkins,
Four little, five little, six little pumpkins,
Seven little, eight little, nine little pumpkins,
Ten little pumpkins now.

Adapted Traditional

Counting Throughout the Day

Provide opportunities throughout the day for your children to help you count. For example, ask them to count as they pass out papers or snacks. Or let them help you count supplies to see which items need to be replaced. Or turn cleanup time after a messy art project into a counting experience by asking each child to pick up a certain number of items from the table or floor and place them in the trash can.

Egg Carton Counter

Punch a hole in the bottom of each egg cup in an empty egg carton. Close the lid and turn the carton upside down so that the holes are facing up. Give a child 12 slot-type clothespins. Have the child put the clothespins in the holes in the egg carton, counting as he or she does so. Remove some of the clothespins and ask the child to count how many are left. Repeat, putting in or taking out clothespins each time.

Counting Cards

Look around the room for familiar objects of which you have 10 or fewer (teddy bears, clocks, tables, easels, balls, etc.). Draw a picture of each object on a separate index card and put the cards together in a deck. Have your children sit in a group. Let each child have a turn drawing a card and naming the object pictured on it. Then have the child walk around counting the number of those objects he or she sees in the room. If desired, have the rest of the children help with the counting or let the children work in teams rather than individually.

Button Cards

For this activity you will need 55 small buttons. Select 10 index cards (or cut 10 rectangles out of cardboard). Sew one button on one card, two buttons on a second card, three buttons on a third card, and so on. Place the cards on a table or on the floor. Let your children touch the buttons as they count how many are on each card. Then have them line up the cards in order from 1 to 10.

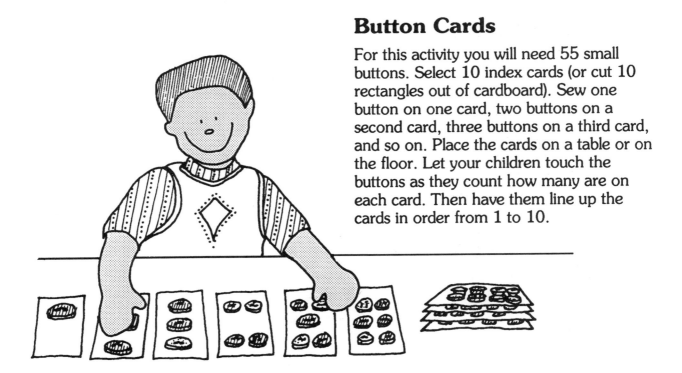

Number Squares

Keep a stack of 10 carpet squares on hand. Encourage your children to count the squares whenever they set them out for various projects. Or place the carpet squares in a line on the floor. Let the children take turns walking or jumping from one square to the next, counting as they go.

Block Towers

Give a child 10 small square blocks to use for building a tower. Encourage the child to count as he or she stacks one block on top of another. Or have several children sit in a circle with the blocks in the middle. Let each child in turn add a block to the tower as everyone counts together.

Counting Sets

For each child cut a large 2 shape out of posterboard. Set out glue and a collection of small objects such as paper clips, rubber bands, pasta shapes, cotton balls, short yarn segments and dried beans. Let your children choose pairs of the small objects and glue them on their posterboard numerals in sets of twos. When the glue has dried, display the numerals and count the glued-on sets with the group.

Extension: Adapt the activity for counting sets of other numbers, if desired.

Fishing Game

Cut 10 small fish shapes out of plastic foam food trays or cardboard milk cartons. Attach a metal paper clip to each fish shape. Then float the fish in a tub of water. Tie a small magnet to one end of a piece of string. Let a child use the string as a fishing line by holding onto the other end (a pole is not necessary). Help the child count the fish as he or she catches them and again when all the fish have been caught. Or do this activity with several children at a time and have everyone count together as each child takes a turn.

Variation: Cut the fish shapes from construction paper, attach paper clips and place the shapes on the floor. Use a wooden spoon or a cardboard paper towel tube for a fishing pole and tie a fishing line, with a magnet attached, to one end.

Matching

- ◆ **Match-Ups**
- ◆ **Pairing**
- ◆ **One-to-One Correspondence**
- ◆ **Matching Sets by Counting**

Crayon Matching

Select two identical boxes of large color crayons. Empty the crayons into a shallow container and mix them up. Place the container on a table. Let your children take turns finding the matching colored crayons in the container and lining them up on the table in pairs.

Variation: Select one box of large crayons. Draw outlines of the crayons on index cards and fill in each outline with a different colored crayon. Let your children place the crayons on top of the matching colored crayon shapes on the index cards.

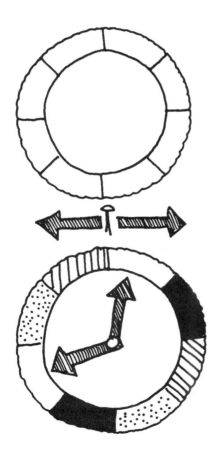

Arrow Match-Ups

Using a black felt-tip marker, divide the rim of a paper plate into eight sections as shown in the illustration. Color each of the four sections on the left-hand side of the plate a different color. Color the four sections on the right-hand side of the plate with matching colors, but in a different order. From stiff paper cut two arrow shapes, as shown, and attach them to the center of the plate with a brass paper fastener. To play, let a child move the arrows around until they point to two colors that match.

Color Lotto

Divide a 9-inch square of posterboard into 3-inch squares. Cut nine 3-inch squares out of different colors of construction paper and glue the squares on the gameboard. Cut matching 3-inch squares out of construction paper. Back the squares with heavy paper to make game cards. To play, have your children take turns placing the game cards on top of the matching colored squares on the gameboard.

Object Outlines

Cut a large rectangle out of sturdy cardboard. Collect several different shaped objects such as a pair of blunt scissors, a pencil, a key, a jar lid, a wooden block and a craft stick. Using a black felt-tip marker, trace around each object on the cardboard rectangle to create a gameboard. Cover the rectangle with clear self-stick paper, if desired. Then give your children the collec-

tion of objects and let them take turns placing each one on top of its matching outline on the gameboard.

Variation: Trace around each object on a separate cardboard square. Set out the objects and the squares and let your children find the match-ups.

Worm Hook-Ups

Cut a large worm shape out of construction paper. Use a felt- tip marker to draw on facial features and to divide the worm into six sections. Draw a different basic shape (a circle, a square, a triangle, a star, etc.) on one side of each dividing line and a matching shape on the other side. Cover the worm shape with clear self-stick paper and cut it into sections along the dividing lines. Let your children piece the worm together by matching the shapes on the ends of the sections.

Craft Stick Match-Ups

Cut large squares out of posterboard. On each square trace around a craft stick to make a different shape, such as a square, a triangle, a rectangle or a diamond. Spread the squares out on the floor and set out a box of craft sticks. Let your children create matching shapes by placing craft sticks on top of the tracings on the squares.

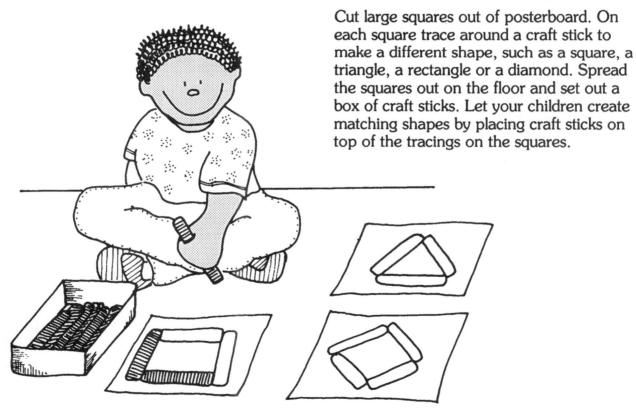

Vegetable Label Match-Ups

Save picture labels from cans or packages of familiar vegetables (tomatoes, corn, beans, peas, carrots, etc.). When you have five or more matching pairs of different vegetable labels, trim the labels with scissors and glue them on separate index cards. Mix up the cards and place them in a box. Let your children take turns sorting through the cards to find the matching pairs of labels.

Butterfly Wing Match-Ups

Collect six cardboard toilet tissue tubes to use for butterfly bodies. Cut 3-inch slits in opposite sides of each tube. Make six pairs of butterfly wings by cutting three paper plates into four sections each. Draw matching sets of dots on each pair of wings. Let your children take turns putting the butterflies together by finding the matching pairs of wings and inserting them into the slits in the cardboard tube bodies.

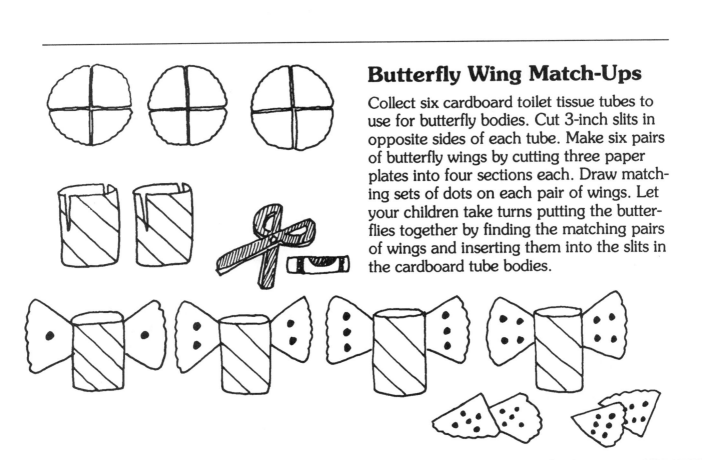

Patterned Egg Match-Ups

Cut 6 to 10 egg shapes out of cardboard. Cover each shape by gluing on a different pattern of wallpaper, wrapping paper or fabric. When the glue has dried, cut each egg shape in half. Mix up the halves and give them to your children. Let them take turns putting the egg shapes back together by matching the patterns on the egg halves.

Picture Match-Ups

Select wrapping paper that contains a pattern of pictures. Make a gameboard by cutting a large rectangle out of the wrapping paper, backing it with cardboard and covering it with clear self-stick paper. From an identical piece of wrapping paper, cut out five or six of the pictures. Cover them on both sides with clear self-stick paper and trim around the edges. To play, let your children take turns placing the cut-out pictures on top of the matching pictures on the gameboard.

Robot Eye Match-Ups

Select a piece of cardboard (about 8½ by 11 inches) to use for making a robot face. With a craft knife, cut two eye holes out of the cardboard as shown in the illustration. Add other details as desired with felt-tip markers. Cut two circles (about 5½ inches in diameter) from posterboard. Divide each circle into four sections. Color the sections of each circle red, yellow, blue and green (or draw different patterns in the sections). Use a brass paper fastener to attach each circle to the cardboard robot face as shown. Let your children turn the circles to make different matching colors appear through the robot's eye holes.

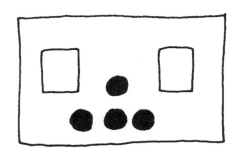

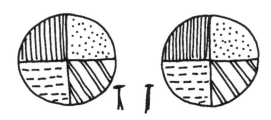

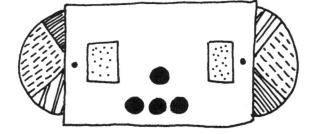

Color Concentration

Cut two squares each from four different colors of construction paper and glue the squares on eight index cards. Mix up the cards and spread them out face down on a table or on the floor. Let one child begin by turning up two cards. If the colors match, let the child keep the cards. If the colors don't match, have the child replace both cards face down exactly where they were before. Continue the game until all the cards have been matched. Let the child who ended up with the most cards have the first turn when you start the game again.

Variation: Adapt the game to match pictures, basic shapes, numerals, etc.

Mother and Baby Animals

Divide 10 index cards into two sets. On one set, draw or glue pictures of different adult animals (a bear, a dog, a fish, a squirrel, a bird, etc.). On the other set, draw or glue pictures of matching baby animals. Mix up the cards and let your children take turns pairing the pictures of the baby animals with their "mothers."

Variation: Instead of drawing baby animals on the second set of cards, draw pictures of animal homes (a cave for a bear, a dog house for a dog, a pond for a fish, a tree for a squirrel, a nest for a bird, etc.).

Three Bears Matching Game

Make felt cutouts to use on a flannelboard when telling the story of the Three Bears. Include shapes for the Father Bear, Mother Bear and Baby Bear. Also include a bowl shape, a chair shape and a bed shape in the proper size for each character. When your children are familiar with the story, let them group the bear shapes with their matching bowl, chair and bed shapes on the flannelboard.

Whose Shoes?

Cut out five or six magazine pictures of people who are wearing shoes. Glue the pictures on cards cut from posterboard or stiff paper. When the glue has dried, cut each card into two pieces just above the shoes that the pictured person is wearing. Mix up the pieces and place them on a table. Let your children take turns putting the cards back together by matching the people with their shoes.

Mitten Mates

Cut a mitten pattern out of cardboard. Use the pattern to make eight or more pairs of mitten shapes from different colored, patterned or textured fabric. Place the shapes in a box and mix them up. Let your children take turns searching through the mitten shapes to find the matching pairs.

Variation: Use real mittens instead of mitten shapes. Give each child a mitten to wear and put all the mates in a pile. Let your children look through the pile to find their mitten mates.

Table Setting Game

Have a child stand with you beside a table. Ask the child to pretend that it is time for dinner and that you need to set the table for four people. Place three or five plates on the table. Ask, "Do I have too many plates or not enough plates?" Continue by asking, "How many more plates do I need?" or "How many plates should I take away?" When four plates have been arranged in place on the table, continue the game using cups, spoons, forks and napkins.

Helpers

To reinforce the concept of one-to-one correspondence, let helpers do tasks such as the following:

- Distribute paintbrushes (or other art supplies) by placing one in front of each chair at the art table.

- Hand one cookie or one box of juice to each child at snacktime.

- Pass out a paper towel to each person who has just finished washing his or her hands.

- Place one copy of a notice to be taken home in each child's cubby.

Chair Antics

Set out five chairs and ask your children to place different objects on each one. For example, give them five teddy bears and ask them to place one bear on each chair. Next, ask them to place a book in each teddy bear's lap. Then ask them to place one hat on each of the teddy bears sitting in the chairs. Continue the game using other kinds of small objects or toys.

Valentine Fun

Around Valentine's Day let each child make a simple valentine for each of the other children in the group. Hand out paper sacks or large envelopes. Let your children decorate their sacks or envelopes to make valentine holders. Help the children write their names on their valentine holders and have them place their holders on a table. Then let each child pass out the valentines that he or she made by placing one in each of the other children's holders.

Variation: Let your children fill Christmas stockings with small treats they have made or Easter baskets with eggs they have decorated.

Bee Stripes

Cut 10 bee shapes from construction paper. Divide the shapes into five pairs. On each pair draw one, two, three, four or five stripes. Cover the bee shapes with clear self-stick paper, if desired. Mix up the shapes and place them on a table. Challenge your children to find each pair of bees by counting and matching the stripes.

Number Cans

Collect five empty juice cans. Cover the cans with construction paper or colored self-stick paper. Number the cans from one to five by drawing on dots or attaching dot stickers. Set out the cans along with 15 craft sticks. Let your children count the numbers of dots on the cans and place matching numbers of craft sticks inside them.

Variation: Instead of craft sticks, use crayons or plastic straws.

Domino Card Game

Making the Cards: Divide each of 21 small index cards in half with a line. Choose six numbers such as zero, one, two, three, four and five. Draw dots to represent those numbers on the halves of the 21 cards (for zero leave the cards blank). Use these combinations: zero-zero, zero-one, zero-two, zero- three, zero-four, zero-five; one-one, one-two, one-three, one-four, one-five; two-two, two-three, two-four, two-five; three-three, three-four, three-five; four-four, four-five; five-five.

Playing the Game: Give each child several domino cards and place the rest face down in a pile. Let one child begin by placing a card in the middle of the playing area. If the next child has a card with a half that matches one of the halves of the first card, have the child place the card next to the first card (either vertically or horizontally) so that the matching halves are touching.

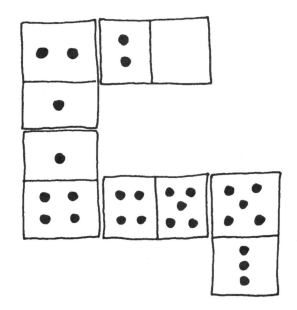

If the child does not have a matching card, let him or her draw cards from the pile until a match is found. Continue the game, letting your children match the cards as they wish, until all the cards have been played.

Block Trains

Set out two piles of small blocks. Ask your children if they think that there are the same number of blocks in each pile. Have them test their responses by lining up the blocks in the piles to create two "trains," one standing next to the other. Then have them count the numbers of blocks in the two trains to see if they match.

Variation: Have your children make block towers instead of trains.

Developing
Shape Recognition

Shape Pictures

Cut various sizes of circles, squares, triangles and rectangles out of lightweight cardboard or posterboard. Store the shapes in a large envelope. Have your children lay out the shapes on a table. Encourage them to move the shapes around in different ways to create pictures or designs.

Variation: Back the shapes with small magnets and let your children manipulate them on a cookie sheet or other metallic surface.

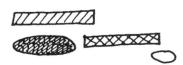

Shape Crowns

Make a crown for each child out of construction paper. From other colors and kinds of paper, cut out small shapes such as circles, squares, triangles and diamonds. Give your children their crowns and let them glue on the small colored shapes to represent jewels.

Shapes Poem

Cut a circle, a triangle, a square and a rectangle out of felt. Place the shapes on a flannelboard, one at a time, as you read the following poem.

The circle is a shape
That's easy to be found.
It has no corners on it
And is completely round.

The triangle is a simple shape,
I think you will agree.
Count its sides and corners
And you will find just three.

Here's a shape that you should know,
A square is its name.
It has four corners and four sides
That measure all the same.

This shape is called a rectangle,
It's found most everywhere.
It has four corners and four sides,
Sort of like a square.

But if you call this shape a square,
You would be very wrong.
Since two of its sides are short
And two of them are long.

Author Unknown

Five Big Cookies

Cut the following shapes out of felt and decorate them to resemble cookies: a square, a circle, a triangle, a rectangle and a star. Let your children help place the shapes on a flannelboard as you read the poem below.

Five big cookies with frosting galore,
Mother ate the square one,
    then there were four.
Four big cookies, two and two, you see,
Father ate the round one,
    then there were three.
Three big cookies, but before I knew,
Sister ate the triangular one,
    then there were two.
Two big cookies, oh what fun!
Brother ate the rectangular one,
    then there was one.
One big cookie, see how fast I run —
I ate the star and now there are none!

Jean Warren

Shape Walk

Use pieces of masking tape to make large outlines on the floor of a circle, a square, a triangle and a rectangle. Let your children take turns walking, crawling or hopping around the edges of the shapes. Or ask each child in turn to identify a particular shape and then walk around it.

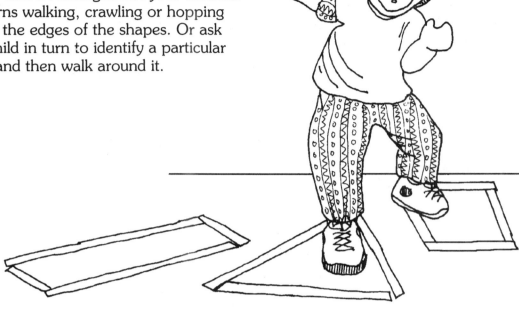

Beanbag Game

Find a sturdy cardboard box. In the bottom, cut out several large shapes such as a circle, a square and a triangle. Place the box upside down on the floor and give a beanbag to a child. Ask the child to try tossing the beanbag into the box through the circle, through the square and then through the triangle.

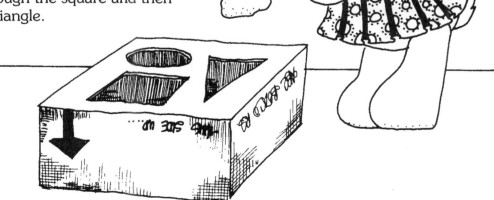

Shapes for Sorting

Choose three colors of construction paper such as red, yellow and blue. From the red paper, cut out a large circle, a medium square and a small triangle. From the yellow paper cut out a large square, a medium triangle and a small circle. From the blue paper cut out a large triangle, a medium circle and a small square. Mix up the shapes and give them to your children. Let them sort the shapes into piles by color, by size and then by shape.

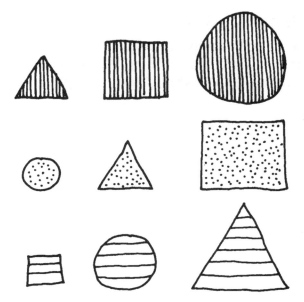

Shape Puzzles

Select three large pieces of thick cardboard. Using a craft knife, cut a circle out of the first piece, a square out of the second piece and a triangle out of the third piece. Make a handle for each cut-out shape by winding a short piece of yarn around a brass paper fastener, inserting the fastener through the center of the shape and bending back the prongs. Set out the cardboard pieces and the shape cutouts. Let your children take turns fitting the cutouts into the matching shaped holes.

Shape Picture Cards

Select twelve index cards. On four of the cards draw pictures of triangular objects (a clown hat, an ice cream cone, a fir tree, a coat hanger. etc.). On four other cards draw pictures of circular objects (a face, a clock, a balloon, a ball, etc.) On the four remaining cards draw pictures of square objects (a box, an alphabet block, an envelope, a building, etc.). Sit together with your children. Lay out three cards containing matching shaped pictures and one card containing a picture that is shaped differently. Ask the children to identify the picture that is different and to tell why it doesn't belong with the other three. Continue, using other combinations of cards.

Shape Formations

On a separate piece of paper for each child, draw simple geometric shapes (a circle, a square, a triangle, etc.), leaving off one side of each shape. Give the papers to your children and let them complete the shapes by drawing lines with crayons or felt-tip markers.

Variation: For a different kind of shape formation activity, pound nails into a board to make outlines of various shapes. Let your children stretch rubber bands around the nails.

Shape Folder

On the insides of a file folder, draw eight basic shapes (a circle, a triangle, a star, a heart, etc.). Draw matching shapes on posterboard and cut them out. Place the cutouts and the open folder on a table. Let your the children match the shapes by placing the cutouts on top of the corresponding shapes on the file folder. Store the cutouts in an envelope taped to the back of the folder, if desired.

Shape Day

Every now and then, make plans with your children to have a Shape Day. Choose a shape such as a circle. Ask the children to bring in circular objects to display on a round table. Or ask them to wear clothes that contain circular designs. Plan activities that center around the circle shape. For example, have the children print circles on paper, count buttons, play with wheeled toys and snack on round crackers.

Developing Number Recognition

◆ **Learning Numerals**
◆ **Writing Numerals**
◆ **Matching Sets With Numerals**

Number Books

Give each child five sheets of paper that have been numbered from 1 to 5, along with 15 colored circle stickers. Hold up a piece of paper with the numeral 1 written on it. First, ask your children to hold up their papers that contain the numeral 1. Next, ask them each to hold up one circle sticker. Then have them attach their circle stickers to their papers. Follow the same procedure to help the children attach matching numbers of stickers to their remaining papers. When you have finished, staple each child's papers together with a colored paper cover to make a book. Print "My Number Book" and the child's name on the front.

Extension: If desired, let your children use crayons or felt-tip markers to turn the circle stickers in their books into such things as balloons, flowers or wheels.

Sandpaper Numerals

Cut numeral shapes out of sandpaper and glue them on separate index cards. Use the cards for activities such as the ones below.

- Place the cards face down on a table. Let each child have a turn drawing a card, touching and naming the numeral on it, then hopping, clapping, etc., that many times.

- Place the cards in a paper bag. Let your children take turns reaching into the bag and identifying the numerals on the cards by touch.

- Attach the cards to a tabletop with masking tape. Have your children place sheets of paper on top of the numerals and color over them with crayons to make rubbings.

Shoelace Numerals

Select several sheets of construction paper and write one large numeral on each sheet. Set out 20-inch-long shoelaces (or pieces of string). Invite your children to create numerals by placing the shoelaces on top of the written numerals on the construction paper sheets.

Variation: Have your children use playdough snakes instead of shoelaces.

Decorated Numerals

For each child cut the numerals 1 to 5 out of posterboard. Set out glue and collections of small objects such as buttons, dried flowers, toothpicks, cotton balls and circle stickers. Help your children glue matching numbers of small objects on their posterboard numerals. For example, have them glue one toothpick on their 1 shapes, two buttons on their 2 shapes, three dried flowers on their 3 shapes, and so on. Display the decorated numerals around the room, if desired.

Numeral Collages

Look in newspapers, magazines and catalogs for the numerals 1 to 10. Cut or tear out as many examples of each numeral as you can find and place them on the art table. Give each child some glue and a piece of construction paper. Let your children choose numerals and glue them any way they wish on their papers to create collages.

Number Day

When you have been working on a number, such as three, make plans with your children for a Number Day. Ask them each to bring in an object that has a 3 on it (a cereal box, a magazine cover, etc.) Also encourage them to wear items of clothing that contain 3's. For other activities, you might wish to read "The Three Bears," serve shamrock cookies and have the children make crowns decorated with 3's.

Touch and Tell

Cut several large numeral shapes out of cardboard. Place the shapes on a table and invite a child to pick them up and name them. Then have the child close his or her eyes. Mix up the shapes and ask the child to try identifying each numeral by touch.

Variation: Cover the numeral shapes with a textured material such as sandpaper, wallpaper or fabric.

Pass the Numerals

Ask your children to sit in a circle. Hand out two or three cardboard numerals. Play some music and have the children pass the numerals around the circle. Whenever you stop the music, ask each child holding a numeral to show it to the group and name it. Continue the game, substituting or adding other numerals as desired.

Erase a Numeral

Using chalk, cover a chalkboard (or a slate) with 1's, 2's, 3's, 4's and 5's. Let each child have a turn finding a numeral on the chalkboard that he or she knows, naming it and then erasing it. Continue until all the numerals have been erased.

Variation: Ask individual children to find specific numerals.

Fishing for Numerals

Cut numeral shapes out of stiff paper and glue a metal paper clip to each one. Make a fishing pole by tying one end of a piece of string to a short dowel and attaching a small magnet to the other end of the string. Spread the numerals out on the floor. Give a child the fishing pole and invite him or her to "go fishing" for numerals. Have the child name all the "fish" that he or she catches.

Variation: Ask individual children to fish for specific numerals.

Spinner Game

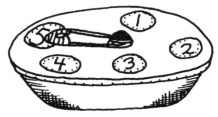

Find an empty whipped topping tub with an umarked plastic lid. Use a brass paper fastener to attach a large closed safety pin to the center of the lid, as shown in the illustration. Adjust the paper fastener so that the safety pin will spin around easily. Attach stickers marked with numerals around the edge of the lid as shown. Let a child have fun spinning the safety pin spinner and naming the numerals to which it points when it stops moving.

Variation: Fill the tub with buttons (or other small objects). After spinning the spinner and naming a numeral, have the child count out that number of buttons.

Zero

To help your children understand the concept of zero, do activities such as the following:

- When asking your children to count objects around them, name a few objects of which there are none. Have the children tell you that there are zero crocodiles, etc., in the room. Then write the numeral 0 where the children can see it.

- At the end of the day, have all your children with brown hair line up. Count the number of children with the group. Next, ask all the children with purple hair to line up. Have the group tell you that there are zero children with purple hair.

- Cut the numerals 0 through 5 plus five seasonal shapes (apples, stars, etc.) out of felt. Form a long piece of yarn into a circle on a flannelboard. Inside the circle place from zero to five shapes. Ask your children to place the matching numeral next to the circle.

Tracing Over Numerals

Cut numerals out of sandpaper and glue them on cardboard squares. Give one of the squares to a child. Have him or her "write" the numeral by tracing over it with a finger. As the child does so, encourage him or her to say the numeral's name. Continue in the same manner using the rest of the numeral squares. Occasionally, ask the child to close his or her eyes and try to imagine a numeral's shape while tracing over it.

Variation: Let the child trace over the sandpaper numerals with a piece of chalk or a cinnamon stick.

Writing Numerals in Sand

Place a layer of sand (or salt) in the bottom of a sturdy shallow box. Let your children take turns writing numerals in the sand with a finger or a craft stick. Show them how to erase their numerals by gently shaking the box back and forth.

Variation: Let your children write in wet sand, either indoors or at the beach.

Yarn Numerals

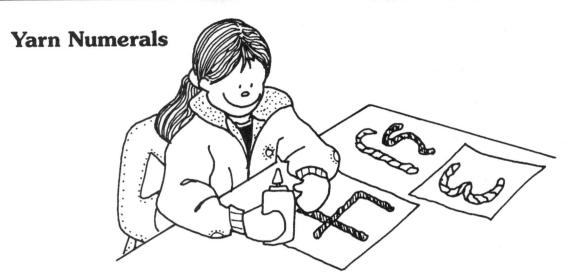

Give each child several pieces of yarn (about 8 inches long). Let your children fashion their yarn pieces into different numeral shapes. Or let them glue their yarn pieces over numerals that have been written on pieces of construction paper. For a textured experience, have the children dip their yarn pieces into liquid starch before arranging them in numeral shapes on sheets of paper. Later, let the children trace over their starched yarn numerals with their fingers.

Variation: Use string or cooked spaghetti instead of yarn.

Numeral Slate

Write the numerals 1 to 10 on a large piece of white construction paper. Cover the paper with clear self-stick paper to make a "slate." Set out the slate along with a black crayon and a few paper towels. Let your children take turns writing over the numerals on the slate with the crayon. Show them how to rub over the slate with a paper towel to erase their crayon marks.

Counting Holes

Select five index cards. On the left-hand side of each card, write a numeral from 1 to 5. Then on the right-hand side, punch a matching number of holes with a hole punch. Let your children take turns counting the numbers of holes in the cards and naming the matching numerals.

Variation: Give older children paper squares marked with numerals and let them punch out matching numbers of holes.

Counting Worms

Cut five apple shapes (about 5 by 5 inches each) out of cardboard. Cut one finger hole in the first shape, two finger holes in the second shape, three finger holes in the third shape, and so on. Color the apple shapes red and mark each one with the numeral that matches the number of holes in it. Let your children take turns choosing an apple shape, sticking their fingers through the holes and then naming the number of "worms" they see.

Counting Fingers

Ask a child to place one hand on a piece of construction paper. Help the child trace around his or her hand with a felt-tip marker. Count with the child the number of fingers on his or her hand tracing. As you do so, write corresponding numerals on the finger shapes. Let the child keep his or her numbered hand shape to use for counting practice.

Counting Hearts

Give each child a paper plate and a plastic sandwich bag containing 10 small paper hearts (or other seasonal shapes). Number 10 index cards from 1 to 10. Have your children sit on the floor in a circle with their paper plates in front of them. Place one of the cards in the middle of the circle and name the numeral on it. Then have the children each count out on their paper plates that number of paper hearts. Have them put their paper hearts back into their plastic bags before you place another numbered card in the center of the circle. Continue the game until all the cards have been played.

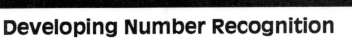

My Number Poster

Help each child create his or her own number poster. To make each one, divide a large piece of heavy paper into 11 rows. Number the rows from 0 to 10 down the left-hand side of the paper. Set out small items such as buttons, paper clips, cotton balls, pasta wheels and circle stickers. Have the child identify a numeral on the chart. Then help the child glue on a corresponding number of small items in a line following the numeral. Continue in the same manner until the poster is complete. If desired, display your children's posters on a wall or a bulletin board to use for counting practice.

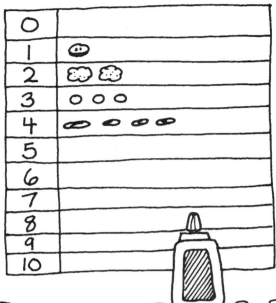

Counting Apples

Cut five squares out of cardboard. On each square glue a leafy tree shape cut from green felt. Write a numeral from 1 to 5 under each tree. Cut 15 small apple shapes out of red felt. Set out the tree cards and the apple shapes. To play, have a child identify the numerals beneath the trees and place matching numbers of apples on the trees.

Circle Clip-Ons

Cut a large circle out of cardboard (or use a paper plate). Divide the circle into six equal sections. Label the sections from one to six by drawing on sets of dots. Write a numeral from 1 to 6 on each of six spring-type clothespins. To play, let your children take turns clipping the clothespins to the matching numbered sections on the cardboard circle.

Variation: Instead of clipping on the clothespins, have your children place them on the circle in the matching numbered sections.

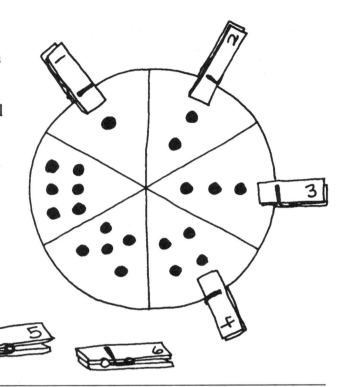

Picture Counting Books

Make a blank book for each child by stapling 10 pieces of white paper together with a colored paper cover. Write "My Counting Book" and the child's name on the front. Number the pages in the book from 1 to 10. Let your children look through magazines or catalogs and tear or cut out small pictures. Then have them glue one picture on the first page of their books, two pictures on the second page, three pictures on the third page, and so on.

Variation: Let your children attach stickers to their book pages.

Make a Set

Select five index cards and number them from 1 to 5. Place the cards in a paper bag. Collect five each of several different objects (blocks, toy cars, craft sticks, etc.). Put each group of objects in a separate container or pile. Let a child begin by taking a card from the bag and naming the numeral on it. Then have the child make a corresponding set by grouping a matching number of identical objects together. Continue the game until each child has had a chance to create several different number sets.

Number Hangers

Tie a clothesline between two chairs at your children's eye level. Label five coat hangers with cards numbered from 1 to 5. Hang the coat hangers on the clothesline in any order desired and set out a basket containing 15 spring-type clothespins.

Invite the children to identify the numerals on the hangers and clip on matching numbers of clothespins. Then encourage them to arrange the hangers on the clothesline in numerical order.

Counting Pan

Number the inside bottoms of six paper baking cups from 1 to 6. Place the baking cups in a 6-cup muffin tin. Give a child a box containing 21 counters (pennies, small buttons, dried lima beans, etc.). Have the child identify the numerals in the bottoms of the paper baking cups and drop in corresponding numbers of counters.

Variation: If paper baking cups are not available, place numbered paper circles in the bottoms of the muffin tin cups.

Paper Clip Game

Select five index cards. Write a numeral from 1 to 5 on each card. Give the cards to a child along with 15 paper clips. Have the child choose one card at a time, name the numeral on it and then attach that number of paper clips to the card.

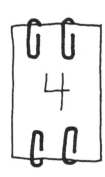

Counting Sticks

Number 10 tongue depressors with numerals from 1 to 10. Draw corresponding sets of dots on 10 more tongue depressors. Place the sticks in two piles and mix them up. Let your children take turns searching through the piles to find the matching numbered sticks.

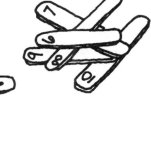

Math Wheel

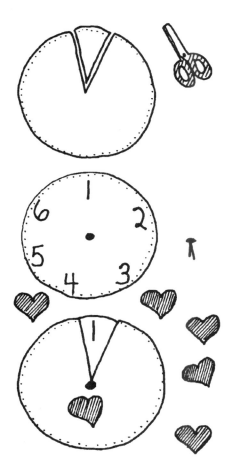

Cut two circles (about 12 inches in diameter) out of cardboard. Cover one circle with felt. Cut a pie-shaped wedge out of the circle as shown in the illustration. Around the edge of the other cardboard circle, write the numerals 1 to 6 in a clockwise fashion. Place the felt-covered circle on top of the numbered one. Poke a hole through the centers of both plates and insert a brass paper fastener. To use the wheel, give a child six small felt cutouts (hearts, bears, stars, etc.). Have the child turn the bottom circle of the wheel until a numeral appears in the wedge-shaped hole. Then ask the child to place that number of felt cutouts on the top circle. Have the child remove the cutouts before turning the math wheel again.

Variation: Have the child place a specific number of shapes on the wheel and then turn it until the corresponding numeral appears.

Tree Planting Game

Paint a shoebox brown and cut six slits in the lid. Number the slits from 1 to 6. Cut six tree shapes out of green construction paper and glue them to craft sticks. Draw from one to six sets of dots on each tree. To play, have your children take turns inserting the trees in the matching numbered slits in the shoebox lid.

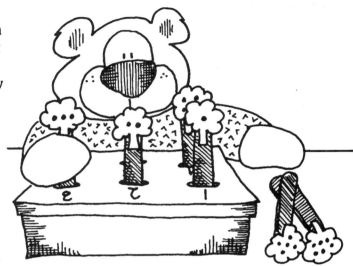

Number Boxes

Collect five cardboard cartons that are approximately the same size. Number the boxes from 1 to 5 by drawing large numerals on the sides. Ask a child to choose a box and identify the numeral on it. Then have the child take the box around the room and fill it with that number of toys, blocks or other objects. If desired, let your children work in pairs and take turns helping each other count out objects into the numbered boxes.

Understanding Relationships

- ◆ **Opposites**
- ◆ **Ordinal Numbers**
- ◆ **Sequence**
- ◆ **Comparisons**
- ◆ **Spatial Relationships**

Learning With Boxes

Collect boxes of various kinds. Use the boxes to introduce common opposites, such as those below, to your children.

- Big — Little
- Open — Closed
- Light — Heavy
- Thick — Thin
- Full — Empty
- Wide — Narrow
- Many — Few
- Far — Near
- First — Last

Hard and Soft

In a bag place objects that are hard (a wooden block, a metal spoon, a stone, etc.) and objects that are soft (a cotton ball, a piece of foam rubber, a woolen sock, etc.). Let one child at a time remove objects from the bag and sort them into two groups: objects that are hard and objects that are soft.

Extension: Place all the objects in a box that has a hole cut in one side. Let each child have a turn reaching through the hole to retrieve something hard or soft from the box.

Left and Right

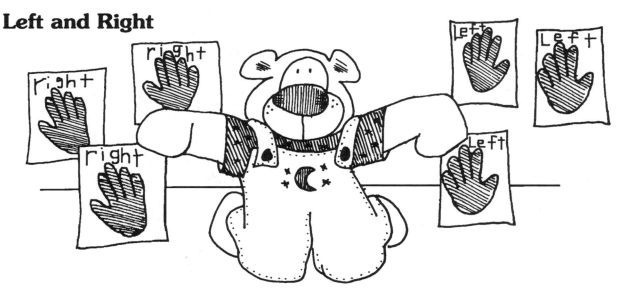

To help your children learn the concepts of left and right, have them do activities that involve their hands. For example, let them make finger rings with red and blue pipe cleaners. Have them wear their red rings on their right hands and their blue rings on their left hands. Or let each child make prints of both hands on a separate piece of posterboard. Label the handprints "Left" and "Right." Display all the handprints on a wall or a bulletin board.

Rough and Smooth

Set out a box of fabric squares. Let your children rub their hands over each square and decide whether it is rough or smooth. Then have them sort the squares into two piles according to texture.

Extension: Let your children go on a hunt around the room. Have them each bring back one object that feels rough and one object that feels smooth. Let each child pass around his or her objects for the others to touch and examine.

Little Train

Have your children line up one behind the other to form a train. Discuss each child's position: "Ashley is the first train car; Joseph is the second train car; Teresa is the third train car," and so on. Then recite the first verse of the poem below for each child and have the child chug across the room. When all the children have crossed the room, have them hook up and chug back together as you recite the second verse of the poem.

(First/second/etc.) little train car
Chugging down the track,
I wonder when
That car will come back.

Here comes the train
Chugging up the track.
I'm so glad
It finally came back!

Jean Warren

Transition Time Learning

Take advantage of transition times to reinforce ordinal concepts. For example, use sentences such as these: "Today, Cody is first in line and Christina is last; When we go outside, Katie's turn on the swing will be first and Adam's turn will be next; This morning, Andrew arrived before Libby, and David arrived after Libby." Also, when talking about an activity that involves several steps, discuss what your children will be doing first, next, and so on.

Animals in the Barns

Cut five barn shapes out of red felt. From other colors of felt cut five different farm animal shapes (a pig, a duck, a cow, a horse, a lamb, etc.). Give the animal shapes to the children and place the barn shapes in a row on a flannelboard. Talk about the positions of the barns. Then let your children place the animals on the flannelboard as you give directions such as these: "Put the pig in the second barn; Put the lamb in the third barn; Put the duck in the first barn."

Extension: When your children become familiar with the game, let them take turns placing different numbers of felt shapes in the barns (three animals in the fifth barn; two animals in the third barn, etc.).

All in a Row

Place five small objects in a box. Invite a child to remove the objects and line them up in any order he or she wishes. Ask the child to tell which object is first in line and which object is fifth, or last. Then ask the child to tell which objects are second, third and fourth. Follow up by asking the child to put the first object, the second object, the third object, and so on, back into the box.

Sequence Cards

On several large index cards draw pictures that show a sequence of events. For example, to illustrate the steps in making a cake, draw pictures of mixing the batter, pouring the batter into pans, baking the cake and frosting the cake. Mix up the cards and place them in a pile. Invite a child to arrange the cards in the proper sequence.

Variation: Draw sequence pictures on small cards and insert them into the sides of separate plastic photo cubes. Let your children arrange the cubes in the proper order. Use sequence ideas such as these: building a snowman; a bird hatching from an egg; the life cycle of a butterfly; the growth stages of a tomato.

Nesting Cans

Make a collection of empty cans that fit one inside the other. For example, include one each of the following sizes: 6-ounce tomato paste can; 12-ounce tomato paste can; 16-ounce fruit can; 29-ounce fruit can. (If an institutional-sized fruit or vegetable can is available, include it also.) Nest the cans one inside the other. Let a child take apart the cans and line them up from smallest to largest or from largest to smallest. Then show the child how nest the cans back together again.

Variation: Instead of cans, use different sizes of boxes.

Numbers in a Row

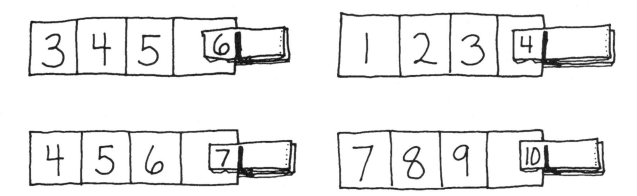

Make numeral strips to use with numbered clothespins for a sequence game. Cut each strip out of stiff paper and use a felt-tip marker to divide it into four sections. Write three consecutive numerals, such as 3, 4 and 5, in the first three sections, leaving the last section blank. Then write the numeral 6 on a spring-type clothespin.

Follow the same procedure to make several more strips using consecutive numerals between 0 and 10. Set out the strips and the clothespins. Let your children take turns clipping the clothespins to the ends of the appropriate strips as shown in the illustration.

Story Cards

Purchase three copies of a small, inexpensive storybook. Keep one copy of the book for reading to your children. Cut out the pictures from the other two copies and glue them on separate posterboard squares. When the children have become familiar with the story, give them the posterboard squares and ask them to arrange the pictures in the proper sequence.

Smallest and Largest

Cut three to five triangles (or other shapes) out of construction paper, making each one slightly larger than the one before it. Mix up the shapes and place them in a pile. Let your children take turns lining up the triangles by size to discover which one is the smallest and which one is the largest.

Variation: Use objects of different sizes such as apples, mixed nuts or stones.

Which Is Bigger?

Invite one child at a time to play this game with you. Name two objects. Then ask the child to tell you which object is bigger. Start with objects like these: a house and a car; an apple and a basketball; a table and a toothbrush; a flower and a tree; a coat and a glove; a pencil and a milk carton. When the child becomes familiar with the game, encourage him or her to ask similar questions for you to answer.

Variation: Name two objects and ask the child to tell you which one is smaller.

Longest and Shortest

Cut ribbon into three to five pieces of various lengths. Mix up the ribbon pieces and place them in a pile. Let your children measure the ribbons one against the other to discover which is the shortest and which is the longest.

Shortest and Tallest

Have two or three children at a time stand together and compare their heights. Which one of them is the tallest? Which one is the shortest? Continue in the same manner, encouraging your children to group themselves in various ways. Finally, have all the children line up together from shortest to tallest.

More Than, Fewer Than

Collect 5 to 10 red poker chips (or use any kind of identical counters). Have your children sit with you in a circle. On a tray in the center of the circle, place three of the poker chips and count them together with the group. Ask the children to close their eyes while you add a poker chip to the tray or take one away. When the children open their eyes, have them count again. Then ask them to tell whether there are more poker chips or fewer poker chips than there were before. Continue playing the game, using different numbers of chips for each round.

Comparison Hunt

In wintertime plan to have an icicle hunt. See who can find the fattest icicle, the thinnest icicle, the longest icicle, the shortest icicle, etc. Or take your children on a nature hike. See who can find the biggest leaf, the tiniest seed, the smoothest twig, the heaviest rock, etc.

Understanding Relationships

Doghouse Game

Make a house for a toy dog by turning a box upside down and cutting a door in one side. Place the doghouse on a table. Put the toy dog in a variety of positions in relation to the doghouse. As you do so, have your children tell you where the dog is (in the doghouse, on top of the dog-house, outside the doghouse, under the doghouse, etc.).

Flower Planting Game

Set out a large box filled with sand. Cut flower shapes out of construction paper and attach them to craft sticks. Let a child plant the flowers in the sandbox as you give directions such as these: "Plant the flowers side by side in two rows; Plant a flower in a corner of the box; Plant one flower next to another flower; Plant one flower behind another flower; Plant two flowers in the middle of the box."

Camera Poses

Give a child a teddy bear (or other object). Using an old camera, pretend to snap pictures of the child as you give him or her directions for different poses. For example, say: "Move backward one step; Take two steps forward; Stand next to the teddy bear; Hold the teddy bear over your head; Put your arms around the teddy bear; Put the teddy bear on your lap; Hold the teddy bear under your arm." Continue in the same manner until every child has had a turn posing for the camera.

Toy Boat Game

Give a child a toy boat to float in a tub of water. Then choose a small plastic toy person and place it in various positions, such as in the boat, under the boat, above the boat, behind the boat and in front of the boat. As you do so, ask the child to tell you where the toy person is in relation to the boat.

I See Something

Have your children sit with you in a group. Start the game by silently choosing an object that you can see nearby, such as a ball. Give a clue such as this: "I see something in a corner of the room that is fun to play with." Have the children try to guess what the object is. As they do so, continue giving clues that describe the object's position. For example, you might say, "It's under a table" or "It's on top of a box, next to a chair." When the children guess correctly, choose another object to describe and start the game again. Or let the children take turns choosing objects and giving clues.

Create a Scene

Arrange a piece of blue yarn across the middle of a flannelboard to represent ocean waves. Cut the following shapes out of felt: a boat, a fish, a whale, a bird, an airplane and a cloud. Let your children arrange the shapes on the flannelboard as you give directions such as these: "Place the boat on top of the water; Place the whale down deep in the water; Place the fish between the boat and the whale; Place the cloud high in the sky; Place the airplane next to the cloud; Place the bird above the boat."

Developing Thinking Skills

◆ **Estimating**
◆ **Observing**
◆ **Following Directions**
◆ **Patterning**
◆ **Thinking Games**

Popcorn Kernel Count

Put some popcorn kernels in a small, clear plastic bottle. Replace the bottle cap, making sure that it is secure. Have your children sit in a circle. Pass around the bottle and ask the children to try guessing how many popcorn kernels are inside. After everyone has had a turn, empty out the kernels and count them with the children to see whose estimate came the closest.

Purse Fill-Up

Give a child a small coin purse that has a zipper. Set out a basket of mixed nuts. Ask the child to estimate how many nuts will fit inside the coin purse. Then count with the child as he or she fills the purse with nuts.

Variation: Have the child fill the coin purse with cotton balls. Ask the child to estimate how many are inside. Then count together as the child empties the purse.

Marble Fun

Place some marbles in a drawstring bag (or tie them in the toe of a large sock). Have your children sit in a circle. Pass around the bag of marbles and let each child try guessing how many are inside. When everyone has had a turn, count the marbles together with the group to see if anyone guessed correctly. Then place a different number of marbles in the bag and start the game again.

Counting Apple Seeds

Hold up an apple and ask your children to estimate the number of seeds that are inside. Write down each child's estimate, if desired. Then cut open the apple, remove the seeds and count them with the children. How close were their estimates to the actual number of seeds? Try the same experiment using a different colored apple.

What Am I Wearing?

Play this game to help your children test their observation skills. Have on hand a large blanket. At an appropriate time, suddenly wrap the blanket around yourself. Then ask the children to try describing the clothes you are wearing.

Look and Tell

Make a game out of observing objects that are similar. For example, let your children take turns naming objects in the room that are a particular color such as red or yellow. Or have them find and name objects that are a particular shape such as round or square. For another kind of observation game, ask the children to look for and name objects that come in sets of two, such as shoes, mittens and salt and pepper shakers.

Developing Thinking Skills

Memory Box

Remove the top from a large cardboard box. Have your children close their eyes while you place the box over an object that sits in a familiar place in the room. When the children open their eyes, have them try guessing what the object is. Let the child who first guesses correctly place the box over a different object when you play the next round of the game.

Bear at the Picnic

Spread a blanket out on the floor. On top of the blanket, place four or five picnic items (an apple, a can of juice, a bag of chips, a paper plate, a spoon, etc.). Choose one child to be the Bear. Have the other children gather around the edge of the blanket and pretend to be napping after eating their picnic lunch. While the children's eyes are closed, have the Bear come and steal away one of the picnic items. When the children open their eyes, let them take turns guessing which item the Bear took. Let the child who first guesses correctly be the new Bear. Then start the game again.

Building Blocks

Give a child six small blocks and a cookie sheet (or a square of heavy cardboard). Ask the child to follow directions such as these: "Place one block in each corner of the cookie sheet; Place two blocks in the middle of the cookie sheet; Make two stacks of blocks, one with two blocks and one with four blocks; Put five blocks in one row; Put six blocks in two rows; Using five blocks, build one short stack and one tall stack."

String Circles

Cut string (or yarn) into 3-foot pieces. Give one piece to each child. Have your children stand in an open area. Then give directions such as these: "Make a circle on the floor with your string; Stand inside your circle; Jump out of your circle; Put one foot inside your circle; Make your circle into a triangle; Now make it into a line."

Map Directions

Draw a road map on a piece of butcher paper. Include pictures of trees, houses, stores, etc. As a child drives a toy car along the roads, give directions such as these: "Turn right and go one block; Turn left by the big apple tree; Go straight ahead for two blocks, then stop at the corner."

Variation: Place a "treasure" on the map where the child can see it. Ask the child to drive the toy car to the treasure. Have the child describe the directions he or she is taking while driving the car.

Four Corners

Give each child a piece of construction paper divided into fourths. Also give each child some glue and four different shapes cut from construction paper (a circle, a triangle, a star, a heart, etc.). Have your children follow directions such as these: "Find the bottom right-hand corner of your paper and place your circle shape there. Place your star shape in the top left-hand corner of your paper." Continue with similar directions as desired. When the game is over, let the children glue their shapes in place.

Poker Chip Patterns

Set out a box containing red and white poker chips. Sit with a child on the floor or at a table. Line up several of the chips in a simple color pattern such as red, white, red, white. Ask the child to copy your pattern using poker chips from the box.

Continue with more patterns, each time making them a little more complex. When the child becomes familiar with the game, encourage him or her to create color patterns for you to follow.

Toothpick Pattern Cards

On each of several large index cards, glue toothpicks in a simple pattern (two vertical, one horizontal, two vertical, one horizontal, etc.). Make each pattern card different. Invite several children at a time to sit with you at a table. Give them each a handful of toothpicks. Let the children select cards and duplicate the patterns on them by lining up their toothpicks on the tabletop.

Developing Thinking Skills

Color Pattern Strips

Select three or more colors of stringing beads and place a number of each color in a box. Cut several strips out of stiff paper. Using felt-tip markers that match the bead colors, draw strings of beads on the strips in different color patterns (red, blue, red, blue; green, yellow, yellow, green, yellow, yellow; etc.). Let your children choose the strips they want and try duplicating the color patterns on them when they are doing bead-stringing activities.

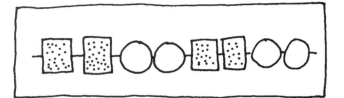

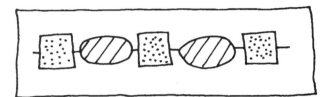

Pegboard Patterns

Make a sturdy gameboard by fitting a piece of pegboard into an old picture frame (check thrift stores or garage sales for inexpensive frames). Set out a box of colored golf tees. Encourage your children to try creating different color patterns as they insert the tees into the holes in the pegboard.

Playing Card Patterns

Find two decks of playing cards. Remove all the hearts from one deck and give them to a child. Then remove all the hearts from the other deck to keep for yourself. On a table or on the floor, line up several cards at a time from your deck. Let the child search for matching cards in his or her deck and lay them out in identical patterns. When the child becomes familiar with the game, take turns creating patterns for the other to follow.

Patterns Everywhere

Encourage your children to look around the room for patterns. Have on hand objects and materials that contain pattern designs. Examples might include curtains, towels, rugs, fabric squares, articles of clothing, jewelry, plates, drinking glasses and wrapping paper.

I'm Thinking of Something

Have your children sit with you in a group. Start by silently choosing a familiar object, such as an apple. Say: "I'm thinking of something round and red that is a fruit." Let the children take turns trying to guess what the object is. If no one guesses correctly, give another clue such as, "It has a core inside that contains small brown seeds." Let the first child to correctly name the object be It for the next round of the game.

What's in the Bag?

Give each child a paper bag in which to hide a small object (a toy, a block, a crayon, etc.). Have your children sit in a circle. Let each child in turn hold up his or her bag and ask the others to try guessing what's inside. Encourage the child to give clues regarding what the object looks like or how it is used. To make the game more challenging, allow only five or six guesses for what is in each child's bag.

Look-Alikes

Select five or more index cards. Using felt-tip markers, draw a clown face on each card, making two of the faces the same and the rest slightly different. Mix up the cards and lay them out on a table. Ask a child to find the two clown faces that are alike.

Extension: Make additional sets of cards by drawing on pictures of such objects as beach balls, trucks, jack-o'-lanterns or decorated Christmas trees.

Look and See

Collect a group of like objects, such as different colored toy cars. Sit with a child and place three of the cars in front of you. Have the child close his or her eyes as you remove one of the cars. When you say, "Look and see," have the child open his or her eyes and tell which of the cars is miss-ing. As the child becomes familiar with the game, start each round with a larger number of toy cars.

Variation: For a more challenging game, use more cars and remove two toy cars at a time or rearrange the remaining cars.

Odd One Out

Have your children sit with you in a group. Explain that you are going to name four things. Ask the children to listen carefully and then tell you which of the things does not belong with the other three. Name things such as these: "Cow, pig, bicycle, chicken; Airplane, pencil, bird, kite; Apple, banana, orange, dog." As the children respond, encourage them to explain their answers.

Variation: Set out three objects that are alike in some way and one object that is different. Ask your children to tell which object is not like the other three. Use objects such as these: a shoe, a hat, a mitten and a block; a plate, a spoon, a cup and a crayon.

What Is It?

Draw pictures of familiar objects on separate index cards. Have your children sit on the floor in a circle. Place the cards face down in the middle of the circle. Let one child begin by choosing a card and holding it up so that the others can see the picture on it but he or she cannot. Have the child try to identify the object pictured on the card by asking the other children questions. If desired, set a limit on the number of questions the child can ask. Continue the game until everyone has had a turn.

Measuring

- ◆ **Linear Measurement**
- ◆ **Weight**
- ◆ **Volume**
- ◆ **Temperature**
- ◆ **Time**

Ribbon Fun

For the measuring activities below, use ribbon that will lie flat when cut into pieces.

- Set out ribbons of different lengths. Have your children find the longest ribbon, then the shortest.

- Cut pairs of ribbons into various lengths. Mix up the ribbons and let your children find the matching pairs.

- Cut ribbons to match the lengths of common objects in the room (tables, pictures, gameboards, books, etc.). Place the ribbons in a box. Let your children select ribbons and search the room to see if they can find objects that are the same lengths as their ribbons.

In Half

Set out a dowel and a piece of string of equal length. Ask your children to imagine that you are going to cut the dowel in half. Can they think of a way that you can determine the exact middle point of the dowel? After the children suggest ideas, show them how to divide the string in half by folding it. Then place the folded string next to the dowel to find the dowel's mid-point.

Straw Game

Cut plastic straws into small pieces of various lengths. Sit with several children at a time. Ask one child to close his or her eyes. Then place a straw section in each of the child's hands and ask the child to tell by touch which section is the shortest and which is the longest. Continue playing the game until everyone has had a turn.

Variation: Cut straw pieces into pairs of matching lengths (one pair for every two children). Give the pieces to your children. Let them walk around the room and compare straw sections to find those that match theirs in length.

Estimating Measurements

Throughout the day, provide opportunities for your children to estimate measurements. For example, you might say, "I wonder how many footsteps long the room is." Then walk the length of the room with the children as you count. Or ask the children to guess how wide a table might be. Then measure it together with such things as hand spans, pieces of ribbon or plastic straws.

Personal Measuring Tools

My Special Measuring Tape — Measure each child's height, then cut a piece of heavy yarn of equal length. Give the child the yarn piece to use as a personal "measuring tape." Encourage your children to use their tapes to measure objects (or other people) around the room. Can they find objects that are the same size as themselves?

My Special Measuring Tools — Trace around each child's hand and foot on posterboard. Cut out each shape, punch a hole in it and attach a shower curtain hook for a handle. Encourage your children to use their special measuring tools by asking questions such as these: "How many hands wide is the sandbox? How many 'feet' long is the rug? Is the rug more or fewer feet long than the table?"

Common Measuring Tools

After your children have had lots of practice in measuring with pieces of ribbon, hand spans, foot cutouts, etc., introduce them to rulers, tape measures and yardsticks. As you encourage the children to experiment with using these linear measuring tools, help them to understand and use terms such as *inch, foot, length, width* and *height.*

Personal Growth

Do the following activities with your children to help them understand how they can measure their own growth.

- Measure each child's height and foot length at the beginning of the year. Repeat 6 to 12 months later. Help the child compare the measurements to discover how much he or she has grown.

- Have your children try putting on clothes that they wore when they were babies. Discuss why the clothes no longer fit.

- Make handprints and footprints of an adult and a baby. Let your children compare them with their own hand and footprints.

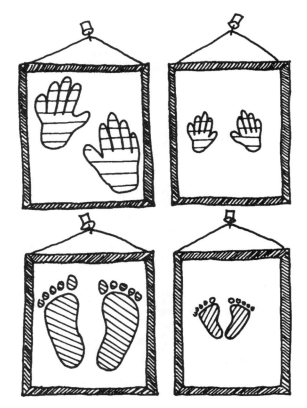

Plant Growth

Let your children help make a "plant ruler" to insert into the soil next to a sprouting plant. On a craft stick draw a line to show how far down into the soil to push the ruler. Then above the line draw marks and numerals indicating 1 inch, 2 inches, 3 inches, and so on. Encourage the children to check the plant's growth each day. If desired, have them record the growth on a height chart as shown in the illustration.

Variation: Have your children draw a picture of the plant. Let them regularly measure the plant to see how many inches it has grown. Then have them add that many inches to the height of the plant in their picture.

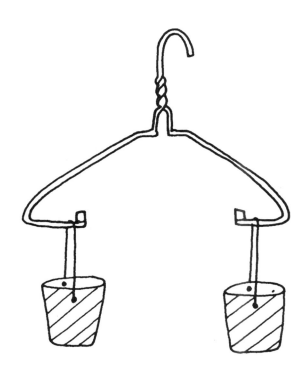

Coat Hanger Balance Scale

Cut the middle section out of the bottom part of a wire coat hanger. Cover the two sharp ends with masking tape and bend them up slightly. Punch two holes in the rims of each of two paper cups. Make a handle for each cup by tying on a 6-inch piece of string. Hang the cups from the cut ends of the coat hanger. To use the scale, place a small object in each cup (try using a penny and a crayon). Balance the coat hanger scale on your finger or a doorknob and ask your children to tell which of the objects is the heaviest. Repeat with other small objects.

Weighing With Scales

Set out bathroom scales and let your children take turns weighing themselves. Help them to keep a record of their weights on a chart. Have them check the chart every few months to see how they are growing in weight.

Extension: Let your children experiment with using other kinds of scales (small postal scales, food portion scales, etc.). Also, talk about how people such as grocers, butchers, doctors and postal employees use scales in their work.

Mini Teeter-Totter

For this activity you will need one half of a log that has been split lengthwise and a long narrow board. Help your children balance the board on the rounded part of the log half to create a mini teeter-totter. Set out blocks of various sizes along with several other objects. Let the children experiment with balancing similar-sized blocks, blocks of different sizes and other objects on their teeter-totter. Encourage them to move the blocks and other objects up and down the board to discover different ways of creating balances.

Estimating Weights

Find four empty margarine tubs that are the same size. Fill each tub with a different kind of material (sand, rocks, cotton balls, etc.), making sure that one of the tubs is noticeably heavier than the other three and that one of tubs is noticeably lighter. Tape the tubs closed and draw or glue on faces, if desired. Let your children take turns lifting the tubs. See if they can identify the heaviest tub and the lightest tub.

Variation: Collect several objects such as an apple, a book, a baby food jar, a pencil and a twig. Have a child close his or her eyes. Then place an object in each of the child's hands. Ask the child to tell which object is lighter or heavier than the other.

Fun With Water

Make a collection of empty cardboard milk cartons that includes one each of the following sizes: half-pint, pint, quart, half-gallon, gallon. During water play let your children experiment with pouring from one carton to another to discover how much water each one holds. (Later, add a plastic measuring cup, if desired.) As the children play, encourage them to use terms such as *full, half-full* and *empty*.

Fun With Sand

Set out a sandbox (or a box filled with cornmeal or rice). Provide a set of plastic measuring cups. Let your children take turns experimenting with measuring the sand. Help them to discover that it takes four quarter-cups or two half-cups to fill the one-cup container.

Extension: Place four 2-ounce containers, two 4-ounce containers and one 8-ounce container on a table. Let your children take turns pouring sand from a pitcher-type measuring cup into the other containers. Encourage the children to estimate and discuss how many of the other containers they can fill with the sand from the one-cup pitcher.

Using a Thermometer

Show your children an outdoor thermometer (or a candy thermometer) and discuss how it is used to measure heat and cold. Place the thermometer in warm water, then in cold water. Have the children observe as the mercury rises and falls. Follow up by letting the children measure the temperatures of other substances such as milk, soup or crushed ice. As you discuss temperature, use terms such as *hot, cold, warm, cool, degrees, Fahrenheit* and *Celsius.*

Extension: Draw a blank thermometer on each of several index cards. Using a red crayon or felt-tip marker, color in the mercury on the thermometers to indicate hot and cold temperatures. Label one box with a summer picture and one box with a winter picture. Let your children sort the thermometer cards into the appropriate boxes.

Play Thermometer

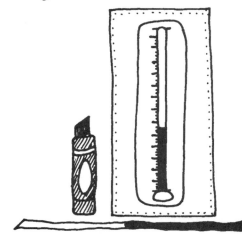

Use a large piece of cardboard and a white ribbon to make a play thermometer. Draw a picture of a blank thermometer on the piece of cardboard. Cut a small slit at the top and bottom of the thermometer and color half of the white ribbon with a red crayon or felt-tip marker. Thread the ribbon through the slits and tie the ends together in back of the cardboard piece. Place a real outdoor thermometer in a spot where your children can easily see it. When they notice a temperature change on the outdoor thermometer during the day, have them change the temperature shown on the play thermometer by moving the red part of the ribbon up or down.

Variation: For each child draw a picture of a blank thermometer on a piece of white construction paper. Cover each picture with clear self-stick paper. Let your children record temperatures on their thermometers by coloring in the mercury with red crayons. To lower the temperatures on their thermometers, have them erase their crayon marks with dry facial tissues.

Holiday Calendars

Use holidays as opportunities to help your children learn about calendars. For example, on the first of December cut a large Christmas tree shape from green felt and 25 circular ornament shapes from other colors of felt. Place the tree on a flannelboard with the ornaments below it. Let your children take turns placing one ornament on the tree each day. Encourage everyone to help you count the number of ornaments left under the tree each day to find out how many days are left before Christmas.

Variation: For an Easter calendar, cut a basket shape and egg shapes out of felt. Or make a chain calendar, as described on page 112, for any holiday desired.

Individual Calendars

At the beginning of the month, give each child a duplicated copy of a calendar page for that month that has a space for a illustration at the top. Let your children use crayons or felt-tip markers to draw seasonal pictures on their calendar pages. Encourage them to check their calendars regularly and to cross out each day after it has passed.

Blank Calendar

Make a large calendar board, as shown in the illustration, that can be used for each month of the year. Screw cup hooks into the board as shown. Write numeral dates on small cards and punch a hole in the top of each card. Have on hand a commercial calendar. Each day, use the commercial calendar to discuss the date and the day of the week. Then let a child hang that day's date card on the appropriate hook on your blank calendar. As your children do this activity, reinforce understanding of the terms *yesterday, today* and *tomorrow*.

Variation: Make the blank calendar on a large piece of posterboard and attach gummed picture hooks on which to hang the date cards.

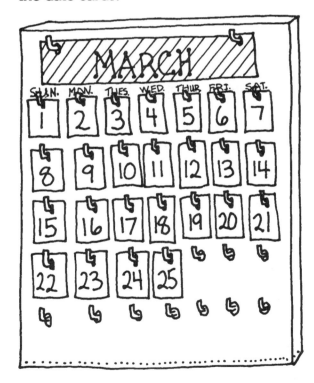

Making a Room Calendar

At the beginning of each month, let your children help make a flip-type room calendar. Select large pieces of heavy paper for the calendar pages (one piece for each day of the month). Number the pages by drawing on large numeral shapes and add the name of the month. Put the pages together, punch holes in the top and insert metal rings. Let your children decorate the pages of the calendar one day at a time, making sure that birthdays, holidays and other special occasions are highlighted. Also, encourage the children to glue on pictures or other materials that relate to activities they are doing that day. At the end of the month, review the calendar pages with the children. Ask them to try recalling some of the things they were doing each day.

Variation: At the end of each month, let your children decorate all the calendar pages for the upcoming month (divide the pages among the children). Display the finished calendar and let the children take turns flipping over the decorated pages each day.

Sand Timer

Select two identical glass jars that have metal lids. Spread strong glue over the two lids and then stick them together top to top. When the glue has dried, use a hammer and a nail to make four medium-sized holes in the lids (pound two holes through one side of the lids and two holes through the other). Fill one of the jars with sand (or salt). Then use the lids to screw the two jars together with the empty jar on top. Tip the jars over and time how long it takes for the sand to stop flowing. Add or remove sand until the timer represents an exact unit of time such as l or 2 minutes. Set out the sand timer at playtime and let your children use it to determine length of turns.

Daily Time Line

Hang up a clothesline for your children next to a wall or a room divider. On separate pieces of paper, draw pictures illustrating the activities you routinely do each day. Cover the pictures on both sides with clear self-stick paper for durability. Each morning let the children help you clip the activity pic-tures to the clothesline in the desired order. (You may wish to color code morning and afternoon activities by attaching self-stick dots.) As the children complete each activity during the day, have them unclip the corresponding picture, turn it around and then reclip it to the clothesline.

Clock Games

Use paper plates (or circles cut from cardboard) to make clocks. Write numerals around the edges of the plates with a black felt-tip marker. Cut a large and small hand for each clock out of heavy black paper and attach them to the center of the plate with a brass paper fastener. Keep the clocks on hand for activities such as these:

- For very young children, remove the minute hand from one of the clocks. Talk about special times such as lunchtime, dinnertime and bedtime. Help the children move the hour hand around the clock to indicate each special time of day.

- Set a paper plate clock for a time when you will be serving the day's snack, reading a story, etc. Have your children watch for the matching time to appear on a real clock that has been placed where the children can easily see it.

- Give each child a paper plate clock. Hold up another paper plate clock that reads 1 o'clock, 2 o'clock, 3 o'clock, etc. Ask your children to make their clocks say a matching time.

Wearing Watches

Make watch faces by cutting circles out of posterboard and writing numerals around the edges. Draw the hands on each watch face to indicate a different hour (1 o'clock, 2 o'clock, 3 o'clock, etc.). Attach the watch faces to elastic bands that will fit around your children's wrists. Have each child select a watch and tell the time shown on it. Then let the child wear the watch as long as desired before exchanging it for one that shows a different time.

Variation: Make pairs of watch faces that show matching times. Let your children put on the watches. Then have them find their "time partners" by walking around and asking one another what times their watches say.

Exploring Other Math Areas

- ◆ **Graphs**
- ◆ **Fractions**
- ◆ **Money**
- ◆ **Addition and Subtraction**

Graphing With Bodies

Let your children use their bodies to create a "graph." Set out signs or pictures that indicate these categories: Brown Eyes, Blue Eyes, Gray Eyes, Green Eyes. Have the children line up behind the signs that best describe their eye colors. Then have the members of each line count off to reveal the number of children who have that eye color. Follow the same procedure to graph such things as color of hair, types of shoes or kinds of pets.

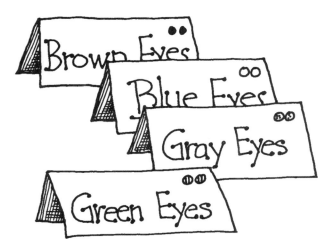

Graphing With Toys

Have each child choose a toy and come join you in a circle on the floor. Make a dividing line in the center of the circle by laying down a piece of string or yarn. Choose a two-part category such as Big Toys and Little Toys. Ask your children to decide whether their toys are large or small. Then have them place the big toys on one side of the dividing line and the little toys on the other side. Continue making new "graphs," using categories such as Hard Toys and Soft Toys or Toys With Moving Parts and Toys Without Moving Parts.

Graphing Shoe Colors

On a piece of butcher paper, draw an outline for a bar graph as shown in the illustration. Using words or pictures, list categories such as the following down the left-hand side of the graph: "Blue Shoes, White Shoes, Red Shoes, Brown Shoes, Green Shoes." Write numerals across the bottom of the graph, as shown, to indicate totals. Count with your children how many of them are wearing blue shoes, white shoes, etc. Then record the numbers by drawing bars on the graph. Discuss the completed graph, asking questions such as these: "Which group is smaller, the one wearing red shoes or the one wearing blue shoes? How many more are wearing brown shoes than blue shoes?"

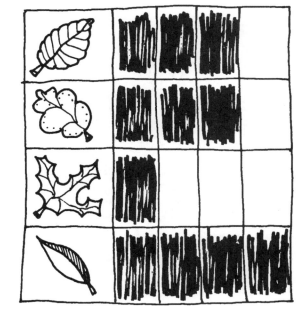

Variation: Let your children help you decide on other topics to record on a bar graph (colors of shirts, numbers of brothers and sisters, favorite foods, kinds of pets, etc.).

Graphing Leaves

Prepare a simple graph by drawing a grid on a large piece of paper. Glue a different shaped leaf in each of the left-hand squares on the grid. Set out matching shaped leaves and let your children sort them into separate piles. Have them count the number of leaves of one shape. Then have them find the matching shaped leaf on the graph and color in one square for each additional leaf. Continue with the rest of the shapes.

Variation: Instead of leaves, use buttons or small seasonal shapes cut from different kinds of materials. Graph the items by color, size, shape, texture, etc.

Circle Fun

Give each child two identical circles cut from one color of paper. Talk about how each shape is a whole circle. Then help each child to fold one of his or her circles in half and to cut along the fold. Have the child put the two halves together to make the circle whole again. Then have the child place the two halves together on top of his or her other whole circle. As your children become familiar with fractions, help them to fold and cut paper circles into fourths, then eighths.

Extension: Cut two round pie shapes out of felt. Decorate the shapes as desired. Cut one pie in half and the other into fourths. Place the pieces on a flannelboard and let your children put them together in various ways to create whole pies.

Heart Halves

For each child cut several heart shapes from different colors of construction paper. Show your children how to fold their hearts in half and help them to cut along the folds. Hand out large pieces of white paper and glue. Have the children identify the two halves of each of their paper hearts. Then have them put their heart halves together and glue them on their papers to create whole hearts.

Puzzle Fun

Select several different colors of posterboard. From each color cut out two identical shapes. Cut one shape in each pair into fractional parts (halves, fourths, etc.). Place the whole shapes on a table. Then invite a child to place the fractional pieces together on top of the matching colored whole shapes. When the child has finished, ask him or her to count the fractional parts of each puzzle.

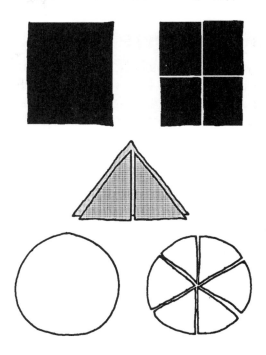

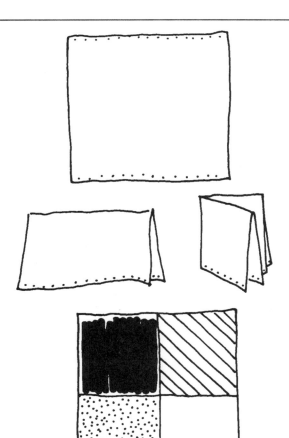

Paper Square Fold

Give each child a square of plain paper. Ask your children to fold their papers in half. Then show them how to fold their papers in half again to create four equal parts, as shown in the illustration. Have the children open their papers and color each fourth a different color. Or let them attach a different kind of sticker in each fourth.

Extension: Later, let your children fold paper squares in half three times to create eighths.

Coin Rubbings

Attach pennies, nickels, dimes and quarters to a tabletop with loops of tape rolled sticky sides out. Let your children place sheets of lightweight paper on top of the coins and rub over them with crayons. When the children have finished, help them write the cents equivalents of the coins next to the rubbings on their papers.

Coin Sorting

In a coin purse place several pennies, nickels, dimes and quarters. Sit with a child and give him or her the purse. Let the child remove the coins one at a time, name them and place them in four separate piles.

Variation: Cut slits in the lids of four small boxes to make "banks." Label the banks by taping on a penny, a nickel, a dime and a quarter. Let the child put matching coins into each bank.

Lemonade Stand

Help your children set up a lemonade stand. Display a sign that reads "Lemonade 3¢." Give the children real coins (or use play money). Let them take turns buying and selling lemonade to one another. Each time they make a money transaction, encourage them to count out the number of coins involved.

Variation: Attach price labels to empty food containers. Arrange the containers on shelves to create a pretend grocery store. Let your children buy and sell food items, using play money.

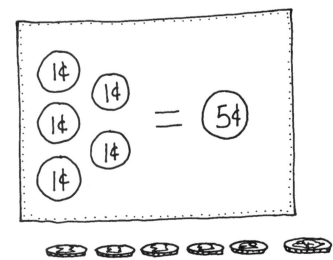

Coin Cards

On a large index card trace around a penny five times. Write "1¢" inside each circle. Draw an equals sign and then trace around a nickel. Write "5¢" inside that circle. Set out five pennies and one nickel. Invite a child to place the coins on top of the circles on the card. Then ask the child to tell you how many pennies equal one nickel. Follow the same procedure using cards showing that two nickels equal one dime, two dimes and one nickel equal one quarter, etc.

Paper Chains

Set out 1- by 8-inch strips of construction paper in seasonal colors. Let your children glue the strips together to create a paper chain for a room decoration. As the children are working, count with them the number of loops in their chain. Ask them to add specific numbers of loops and then count again.

Extension: Let your children make a calendar chain that contains as many loops as there are days remaining before a holiday or other special date. Each day have the children remove one of the loops. Count together the remaining loops to discover how many days are left before the holiday.

Learning Rhymes

Use rhymes like the ones below to develop understanding of adding and subtracting. Let your children raise or lower their fingers (or manipulate felt shapes on a flannelboard) as you recite the rhymes.

Halloween Goblins

One little goblin standing at the door.
Two little goblins dancing 'cross the floor.
Three little goblins peeking through the latch.
Four little goblins, what a happy batch!
Five little goblins, and more that can't be seen.
They're all getting ready for Halloween!

Author Unknown

Five Wee Leprechauns

Five wee leprechauns scurrying by my door,
One jumped away, then there were four.

Four wee leprechauns climbing in my tree,
One hid in the green leaves, then there were three.

Three wee leprechauns, just a busy few,
One went for his pot of gold, then there were two.

Two wee leprechauns having lots of fun,
One hopped over the rainbow, then there was one.

One wee leprechaun with all his work done,
He slipped off for a nap, then there were none.

Susan M. Paprocki

Adding and Subtracting Games

• Place five to seven apples in front of you and have your children count them. Ask the children to close their eyes while you add or remove several apples. When the children open their eyes, help them figure out how many apples have been added or taken away.

• To help your children learn how to subtract by ones, let them practice counting backward from 5 or 10. For extra fun, call out "Blast off!" each time you finish counting.

• Count with your children from 1 to 10. Then ask questions like these: "What number comes before two? What number comes after seven? What number comes after three and before five?"

Beehive Game

Cut a beehive and 5 to 10 bee shapes out of felt and place them on a flannelboard. Move the bees in and out of the hive as you tell a story that involves adding and subtracting. For example, say: "Let's peek into the hive this morning and count the bees. How many are there? Two bees fly outside to look for flowers. How many are left inside the hive? Here come three more bees looking for flowers. How many are outside the hive now?"

Opportunities for Learning Math

- ◆ Snack Time
- ◆ Science Time
- ◆ Outdoor Time
- ◆ Language Time
- ◆ Music Time
- ◆ Art Time
- ◆ Play Center Time
- ◆ Game Time
- ◆ Transition Time

Learning With Apples

Set out a bowl of red, yellow and green apples. Let your children group the apples by color and count the number in each group. Then have them line up all the apples by size to discover which one is the smallest and which one is the largest.

Extension: Choose one of the apples and hold it up. Using the names of your children, say something like this: "Cameron wants to share this apple with Jessica. What can he do?" Help the children to discover that Cameron will need to divide the apple into two parts. Then cut the apple in half. Let the children decide how many more apples will be needed for everyone to receive one half. Then cut that number of apples in two and give the halves to the children.

Learning With Sandwiches

Let your children work with you to make sandwiches for everyone. Help the children cut some of the sandwiches in half to create rectangles. Then help them cut the remaining sandwiches in half to create triangles.

Variation: Give each child a table knife and a slice of bread on a plate. Have your children cut their bread slices into halves, then into fourths. Follow up by letting them spread peanut butter on their bread pieces.

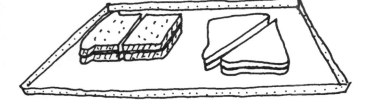

Pizza Fun

When serving a whole pizza you have made for a snack, let your children observe as you first cut it in half. Count together the number of pieces. Continue cutting the pizza in half to create fourths and eights, each time counting the number of pieces with the group. As you place the servings on plates, point out that each piece is one-eighth of the pizza.

Variation: Follow the same procedure when serving a pie.

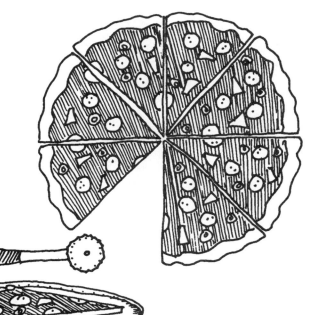

How Many?

Let your children have fun counting food items at snacktime. For example, give them fresh peas to shell. Have them count the number of peas in each pod before they place them in a bowl or pan. Or let each child take a handful of popcorn, peanuts, cereal pieces, etc. Before eating, count together the number of items in each child's hand.

Variation: Set out a bowl of small snack items such as raisins, pretzels or mini-crackers. Let each child in turn roll a die and name the number that comes up. Then help the child count out that number of food items onto his or her plate.

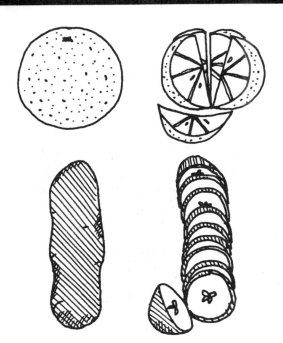

Fruit and Vegetable Puzzles

When cutting up fruits and vegetables for snacks, occasionally demonstrate how the pieces can be put back together to create wholes. For example, divide an orange into segments or slice a zucchini into thick rounds. Let your children observe as you fit the orange or zucchini pieces back together.

Extension: For each child peel a small banana or carrot and cut it into three or four pieces. Mix up the pieces and place them on a plate. Before eating, have the child try fitting the pieces back together to create a whole banana or carrot.

Serving Snacks

Reinforce understanding of one-to-one correspondence and beginning division by letting your children help you pass out snacks. For example, after setting the snack table, count together the number of plates. Set out a tray that contains three crackers for each child. Let the children place one cracker on each plate. Have them count how many crackers are left. Then let them help you decide how many more crackers can be placed on each plate. Follow the same procedure using other snack items.

Opportunities for Learning Math

Measuring and Weighing

Use snacktime as an opportunity for doing measuring and weighing activities. For example, when serving snacks, have your children pour out half-cups or whole cups of juice. Or when helping to prepare snacks, let the children participate in measuring ingredients in teaspoons, tablespoons, cups, pints, quarts, etc. Let them also use kitchen scales to weigh ingredients in ounces and pounds.

Charts and Graphs

Estimate Charts — Make a chart for recording estimates that contains a list of your children's names. Hold up a box of crackers and let each child estimate how many are inside. Help the child write that number on the chart after his or her name. Open the box and count the number of crackers. Then compare the number with the children's estimates. Later, let the children record other estimates such the number of seeds in an apple, the number of cookies in a bag or the number of peaches in a can.

Food Preference Graphs — Divide a long piece of paper into two columns. Choose a food item such as juice. Draw a glass of grape juice at the top of one column and a glass of orange juice at the top of the other. Set out squares of purple and orange construction paper. Let your children show their juice preferences by gluing purple squares in the grape juice column and orange squares in the orange juice column. Ask: "Which juice do people like best? How many more like _____ juice better than _____ juice?" Follow a similar procedure to graph cracker preferences, fruit preferences, etc.

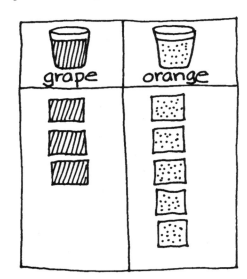

Summer Activities

Shell Fun — Have on hand a wide variety of shells. Let your children sort the shells into groups by kind, size, texture, etc. Have them count the number of shells in each group. Continue by asking the children to find the smallest shell. The largest shell. Then have them take turns balancing a shell in each hand to see which is heaviest and which is lightest.

Counting Flower Petals — Bring in several daisies. Let your children help pull off the petals while everyone counts. Or ask the children to bring in different kinds of flowers. Help them to graph the flowers according to the numbers of petals they have.

Measuring Water — Place a set of plastic measuring cups in a tub of water or a plastic wading pool. Let your children experiment with pouring water from one cup to another. How many half-cups does it take to fill one whole cup? How many quarter-cups?

Floating and Sinking Chart — Collect items that float in water and items that do not. On a piece of paper, draw three columns. Title the first column "Float?" and list the names of the items under the title. Write "Yes" and "No" at the top of the second and third columns. Let your children place each item in a tub of water and record whether or not it floats by making a check mark in the appropriate column on the chart. Count together the number of check marks in each column and discuss with the children the results of their experiment.

Fall Activities

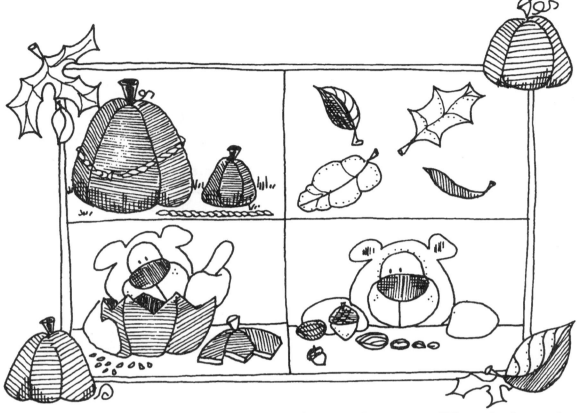

Weighing and Measuring Pumpkins —
Bring in several pumpkins of different
sizes. Let your children weigh the pumpkins on a bathroom scale to discover which
is heaviest and which is lightest. Then let
the children measure the pumpkins with
pieces of yarn that have been cut to fit
around each one. Which pumpkin is the
biggest? The smallest?

Counting Pumpkin Seeds — When
carving a small jack-o'-lantern, cut the top
off the pumpkin and ask your children to
estimate how many seeds there are inside.
Scoop out the seeds, separate them from
the pulp and count them with the group.
Were any estimates too high? Too low?
Just right?

Sorting Leaves — When the leaves begin
to fall, collect several each from four or five
different kinds of trees. Mix up the leaves
and place them on a table. Let your children sort the leaves into like piles and then
count the number of leaves in each one.

Mixed Nut Fun — Set out a bowl containing a variety of unshelled nuts. Let your
children sort and count the nuts by kind.
Or let them line up the nuts from smallest
to largest. If desired, have the children try
weighing the nuts on a balance scale (see
page 96).

Winter Activities

Fun With Ice — Place ice cubes (or icicles) in paper cups. Let your children put one cup outside, one in a refrigerator, one near a heater, etc. Have the children estimate which cubes will melt first and help them keep track of melting times.

Counting Snowflakes — When it is snowing, let your children try catching flakes on pieces of dark colored construction paper and counting them. Help the children discover that each snowflake has six points.

Measuring Snow — Fill several containers of various sizes and shapes with snow. Ask your children to guess which container holds the most snow and which container holds the least. Let the children observe as the snow melts. Then help them measure the water in the different containers to see if their guesses were correct.

Temperature Chart — Place an outdoor thermometer in a place where your children can easily see it. At the beginning of each week make a weather calendar. Help the children record the temperature on the calendar daily.

Spring Activities

Rain Gauge — Use a permanent felt-tip marker to mark off inches on the side of a clear glass jar. Place the jar outside. After a rainfall, check the jar with your children and help them to determine how many inches of rain fell.

Sprouting Seeds — Have your children count as they plant marigold seeds in a large container. When the sprouts first appear, have the children count again. Did all of the seeds sprout? Encourage the children to regularly observe and measure the growth of their plants.

Counting Eggs — Use eggs and an egg carton to help your children understand the meaning of a dozen and a half dozen. Follow up by having the children use the carton to count out a dozen seeds (or other small nature items).

Butterfly Chart — Draw pictures illustrating the life cycle of a butterfly on a large piece of paper. Include pictures of the egg, the caterpillar, the chrysalis and the butterfly. Discuss with your children which stage comes first, next, and so on.

Hop Along

Take your children outside to an open space. Mark off a starting line and a finish line. Have several children at a time stand along the starting line. Let one child begin by spinning a spinner, naming the number that comes up and hopping forward that number of times. Then let the next child have a turn. Continue the game in the same manner until all the players have crossed the finish line.

Number Jump

With chalk draw a line of 10 squares on a sidewalk. Write the numerals 1 to 10 inside the squares, as shown in the illustration. Let your children take turns counting the numbers while jumping from square to square. When the children become familiar with the game, have them take turns following directions such as these: "Jump to 5. Jump forward two squares. Jump back three squares and name the number you are standing on."

Extension: Let your children play Hopscotch.

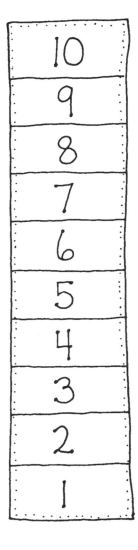

Bottle Bowling

Collect three to five plastic 2-liter bottles. Place about 2 inches of sand in each bottle and replace the caps. Stand the bottles on a sidewalk or other flat surface. Give a medium-sized ball to a child. Let the child try knocking the bottles over by rolling the ball at them. After each try, count with the child how many bottles are down and how many are still standing. When the child has knocked over all the bottles, let the next child have a turn.

Playground Math

- Have your children count ball bounces or pushes in a swing.

- When starting a game, discuss who is going to be first, second, third, and so on.

- Ask your children to divide into two teams, four teams, etc., then count the number of players on each team.

- Encourage your children to try balancing a teeter-totter in various ways.

- Have your children describe how they are climbing up the slide, over the bars, through the tunnel, etc.

- When playing games, let your children use different colored plastic clothespins or other counters to keep score.

- After playtime, encourage your children to sort their toys into the proper bins when putting them away.

Measuring Shadows

On a sunny day, stand a small potted tree (or other appropriate object) on a cement surface. Each hour, use a different colored piece of chalk to outline the tree's shadow. Then help your children measure and compare the different shadow lengths.

Variation: Let your children help outline and measure the lengths of their own shadows.

Sandbox Math

Set out sand toys such as pails, shovels and measuring cups. Let your children count the toys. Then have them distribute one pail and one shovel to each person playing in the sandbox. Encourage the children to compare pails. Are some larger than others? Ask the children to see how many cups of sand it takes to fill the different pails. Which one holds the most sand? The least?

Opportunities for Learning Math

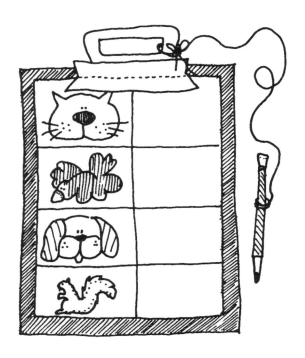

Outdoor Survey

Take your children on a walk to look for different kinds of animals. Give them a survey form attached to a clipboard on which you have drawn pictures of a cat, a dog, a bird, a squirrel, etc. Each time one of the animals is spotted, have the children make a check mark beside the appropriate picture. When you return from your walk, help the children graph the results of their survey.

Variation: Have your children make a survey of different kinds of vehicles or of numerals from 0 to 9 that they *see* on signs and buildings.

Nature Walk

When you go on a nature walk with your children, provide them with bags or boxes to use for collecting nature items. Along the way, look for such things as seeds that come in two parts, three-leaf clovers, leaves that have four points, etc. Later, arrange a time for the children to sort and count the nature items they collected. If desired, let them make a graph by attaching their collected items in like rows on a piece of cardboard.

Reading Children's Books

When reading familiar storybooks to your children, look for opportunities to develop math concepts. For example, try using Eric Carle's *The Very Hungry Caterpillar* in the following ways: Discuss the life cycle of a butterfly to develop understanding of sequence; Talk about the days of the week mentioned in the story to introduce time concepts; Count how many things the caterpillar ate each day and discuss what he ate first, second, and so on, to reinforce counting skills.

Following is a list of other popular children's books that can be used to introduce and review math concepts.

- *Bunches and Bunches of Bunnies* by Louise Mathews (Scholastic). Through a rhyming text and charming pictures, we see bunnies multiply up to 12 times 12.

- *Counting Wildflowers* by Bruce McMillan (Morrow). From 1 to 20 common wildflowers are pictured in colorful photographs.

- *Mouse Count* by Ellen Walsh (Harcourt). A snake wants to capture in a jar as many mice as he can eat, but he becomes too greedy and the mice all escape.

- *Over in the Meadow* by John Langstaff (Harcourt). An old counting rhyme and song that children will want to repeat over and over again.

- *Ten Little Rabbits* by Virginia Grossman (Chronicle). From 1 to 10, we see rabbits wearing various traditional Native American clothes.

- *The Doorbell Rang* by Pat Hutchins (Morrow). Some children start with a dozen cookies but end up having to share them with friends each time the doorbell rings.

- *The Grouchy Ladybug* by Eric Carle (Harper). Unwilling to share aphids with a friendly ladybug, the grouchy one searches from 6 a.m to 6 p.m. for something bigger to eat.

- *Too Many Eggs* by Christina Butler (Godine). We help Mrs. Bear keep track of how many eggs she has put in the birthday cake by moving egg cutouts from page to page.

- *Two Ways to Count to Ten* by Ruby Dee (Henry Holt). One clever animal realizes that you can count to ten faster if you count by twos.

- *What Comes in 2's, 3's and 4's?* by Suzanne Aker (Simon and Schuster). The marvelous photographs in this book show common objects that come in twos (hands), threes (tricycle wheels), and fours (table legs).

Nursery Rhyme Fun

Use nursery rhymes as a fun way to help your children learn math concepts. Following are a few suggestions.

Baa, Baa, Black Sheep — Give each child three paper lunch bags. Ask if the bags are empty or full. Let your children fill their bags with crumpled newspaper pieces. When they each have "three bags full," recite with them the rhyme below. Have the children point to their bags, one at a time, as they say the last four lines.

Baa, baa, black sheep,
Have you any wool?
Yes sir, yes sir,
Three bags full.
One for my master,
And one for my dame,
And one for the little boy
Who lives down the lane.

Traditional

One, Two, Three, Four, Five — Set out several plastic toy animals and invite a child to sit with you. Help the child recite the rhyme below, substituting the name of one of the toy animals for the word *fish*. Have the child "catch" the animal he or she names and then "let it go again." Continue the game using the other plastic animals.

One, two, three, four five,
I caught a fish alive.
Six, seven, eight, nine, ten,
I let it go again.

Traditional

More Nursery Rhymes — Check a Mother Goose collection for other rhymes that can be used for teaching math concepts. Following are a few examples.

- "One, Two, Buckle My Shoe" (Counting)
- "Humpty Dumpty" (Whole and parts)
- "Diddle, Diddle, Dumpling" (Opposites)
- "Hickory, Dickory, Dock" (Time)
- "Old Mother Hubbard" (Empty sets)
- "Thirty Days Hath September" (Months of the year)
- "The House That Jack Built" (Sequence)
- "Pease Porridge Hot" (Temperature)

One Whole Pie

Make a large round pie shape out of felt. Cut the pie into four equal pieces. Put the pieces together on a flannelboard in the shape of a whole pie. Then read the poem below and let your children take turns removing the pie pieces.

One whole pie
Set by the door,
Cut into pieces,
I count four.

Four pieces of pie
All for me,
I ate one piece,
Now there are three.

Three pieces of pie
For me too,
I ate another piece,
Now there are two.

Two pieces of pie
Oh, what fun!
I ate another piece,
Now there is one.

One piece of pie
I can't wait!
I ate that piece,
Empty plate!

Jean Warren

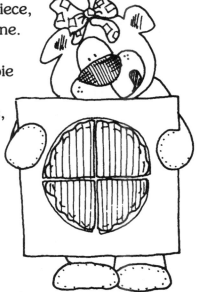

Bo-Peep Counting Rhyme

Cut five lamb shapes out of white felt and place them on a flannelboard. Let your children take turns removing the shapes as you read the poem that follows.

Little Bo-Peep had five little sheep
That played by the cottage door.
One ran away while out at play,
Then Little Bo-Peep had four.

Little Bo-Peep had four little sheep
That played by the old oak tree.
One ran away while out at play,
Then little Bo-Peep had three.

Little Bo-Peep had three little sheep
That played with Little Boy Blue.
One ran away while out at play,
Then Little Bo-Peep had two.

Little Bo-Peep had two little sheep
That played all day in the sun.
One ran away while out at play,
Now Little Bo-Peep has one.

Jean Warren

Kids on the Bed

Place a piece of felt on a flannelboard to represent a bed. Cut five child shapes from other colors of felt and arrange them on the bed shape. As you read the following poem, let your children take turns removing the felt child shapes from the flannelboard.

Five little kids jumping on the bed,
One fell down and bumped his head.
He fell off and rolled out the door.
Kids on the bed? Now there are four.

Four little kids jumping on the bed,
One fell down and bumped her head.
She fell off and bumped her knee.
Kids on the bed? Now there are three.

Three little kids jumping on the bed,
One fell down and bumped his head.
He fell off, he's black and blue.
Kids on the bed? Now there are two.

Two little kids jumping on the bed,
One fell down and bumped her head.
She fell off, no more fun.
Kids on the bed? Now there is one.

One little kid jumping on the bed,
This is what his mother said:
"No more jumping, turn out the light.
Now it's time to say goodnight."

Adapted Traditional

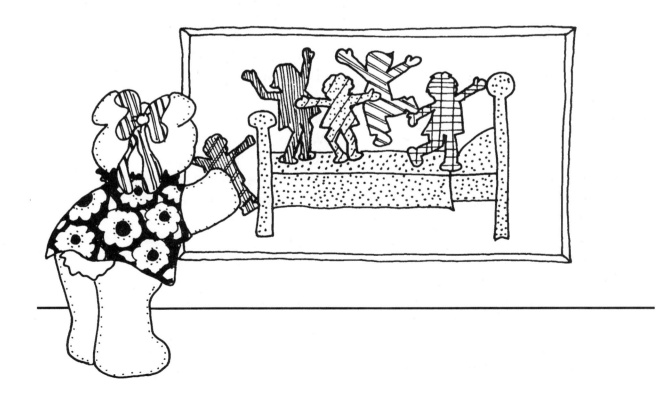

Five Little Frogs

On a flannelboard arrange a piece of yarn in a circle to represent a pond. Cut five frog shapes out of green felt and place them inside the circle. Recite the poem below with your children. Let them take turns filling in the blanks and removing the frog shapes.

Five little frogs
Were down at the pond,
Down at the pond at play.
Along came a hungry _____,
And chased one frog away.

Four little frogs
Were down at the pond,
Down at the pond at play.
Along came a wiggly _____,
And chased one frog away.

Three little frogs
Were down at the pond,
Down at the pond at play.
Along came a giant _____,
And chased one frog away.

Two little frogs
Were down at the pond,
Down at the pond at play.
Along came a purple _____,
And chased one frog away.

One little frog
Was down at the pond,
Down at the pond at play.
Along came a flying _____,
And chased the frog away.

Now no little frogs
Are down at the pond,
Down at the pond at play.
Where do you think the little frogs went
When they all hopped away?

Sue Foster

Ten Red Apples

Cut a tree shape and 10 red apple shapes out of felt. Place the tree on a flannelboard and put the apples on the tree. Manipulate the apple shapes as you read the poem that follows.

Ten red apples growing on a tree,
> (Count apples.)

Five for you and five for me.
> (Count five apples, then five more.)

Help me shake the tree just so,
> (Pretend to shake tree.)

And ten red apples fall down below.
> (Place apples beneath tree while counting.)

1, 2, 3, 4, 5, 6, 7, 8, 9, 10.

Author Unknown

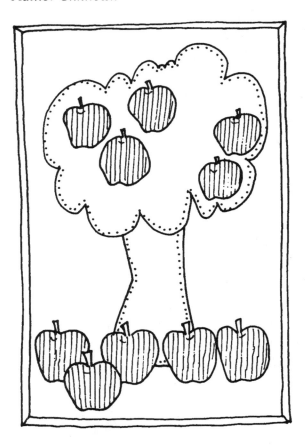

Five Little Flowers

Cut five flower shapes out of felt and place them on a flannelboard. Let your children take turns "picking" the flowers as you read the poem below.

Five little flowers,
Growing outside my door.
I picked one for Grandma,
Now there are four.

Four little flowers,
The prettiest I've seen.
I picked one for Grandpa,
Now there are three.

Three little flowers,
Just a lovely few.
I picked one for Mommy,
Now there are two.

Two little flowers,
Reaching for the sun.
I picked one for Daddy,
Now there is one.

One little flower,
A colorful little hero.
I picked it just for you,
Now there are zero.

Susan M. Paprocki

Four Little Stars

Cut four star shapes out of felt and place them on a flannelboard. Remove one star at a time as you recite the following poem.

Four little stars
Winking at me.
One shot off,
Then there were three.

Three little stars
With nothing to do.
One shot off,
Then there were two.

Two little stars
Afraid of the sun.
One shot off,
Then there was one.

One little star
Alone is no fun.
It shot off,
Then there were none.

Jean Warren

Four Colored Eggs

Cut one egg shape each from the following colors of felt: blue, green, red, yellow. Place the shapes on a flannelboard as you read the poem below.

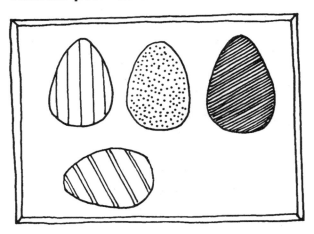

Blue egg, blue egg,
Oh, what fun!
Blue egg, blue egg,
I found one.

Green egg, green egg,
I see you.
Green egg, green egg,
Now I've two.

Red egg, red egg,
Now I see.
Red egg, red egg,
Now I've three.

Yellow egg, yellow egg,
Just one more.
Yellow egg, yellow egg,
Now I've four.

Jean Warren

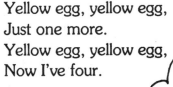

Opportunities for Learning Math

Valentine Friends

Choose five children to stand in a line
and be valentines. Talk about who is first,
who is second, and so on. Then recite
the poem below and have each "valentine" in turn make the appropriate
movement.

Five little valentines,
So bright and gay,
Were waiting to be given
On Valentine's Day.

And just when they thought
That their wait would never end,
Each found a way
To make a special friend.

The first little valentine,
Lacy and pink,
Told someone "I like you"
With a great big wink.

The second little valentine,
Painted white and red,
Told someone "I like you"
With a pat on the head.

The third little valentine,
A flowery miss,
Told someone "I like you"
With a great big kiss.

The fourth little valentine
Cried, "Take me, take me, please!"
And told someone "I like you"
With a great big squeeze.

The fifth little valentine,
The last one in the pile,
Told someone "I like you"
With a great big smile.

Vicki Claybrook

Clap One, Two, Three
Sung to: "Row, Row, Row Your Boat"

Clap, clap, clap your hands,
Clap them one, two, three.
The more you clap, the more we count,
So what will your count be?

One, two, three, four,
Five, six, seven.
The more you clap, the more we count,
Eight, nine, ten, eleven.

Adapted Traditional

I Can Count
Sung to: "Frere Jacques"

I can count, I can count,
One, two, three,
One, two, three.
I can count higher,
I can count higher.
Four, five, six,
Four, five, six.

I can count, I can count,
One, two, three,
Four, five, six.
I can count higher,
I can count higher.
Seven, eight, nine,
Seven, eight, nine.

I can count, I can count,
One, two, three,
Four, five, six.
I can count higher,
I can count higher.
Seven, eight, nine,
Ten, eleven, twelve.

Saundra Winnett

One, Two, Three
Sung to: "This Old Man"

One, two, three, count with me.
It's as easy as can be —
Four, five, six, seven, eight, nine, ten.
Now let's start it once again.

Judy Hall

When the Numbers March Right In
Sung to: "When the Saints Go Marching In"

Oh, when the numbers march right in,
Oh, when the numbers march right in,
We will count them one by one,
When the numbers march right in.

Oh, one, two, three, and four, five, six,
And seven, eight and nine and ten.
When we finish all our numbers,
We will count them once again.

Judy Hall

Sing a Song of Numbers
Sung to: "Sing a Song of Sixpence"

Sing a song of numbers,
Count them one by one.
Sing a song of numbers,
We've only just begun.
One, two, three, four, five, six,
Seven, eight, nine, ten.
When we finish counting them,
We'll start them once again.

Judy Hall

Heigh-Ho, Our Numbers We Know
Sung to: "The Farmer in the Dell"

Number one is here,
Number one is here.
Heigh-ho, our numbers we know!
Number one is here.

Number one takes the two,
Number one takes the two.
Heigh-ho, our numbers we know!
Number one takes the two.

Number two takes the three,
Number two takes the three.
Heigh-ho, our numbers we know!
Number two takes the three.

Continue with similar verses. Let the children wear
numbered necklaces and act out the song as they
would "The Farmer in the Dell."

Lois E. Putnam

Shapes
Sung to: "Frere Jacques"

This is a square, this is a square,
How can you tell? How can you tell?
It has four sides,
All the same size.
It's a square, it's a square.

This is a circle, this is a circle,
How can you tell? How can you tell?
It goes round and round,
No end can be found.
It's a circle, it's a circle.

This is a triangle, this is a triangle,
How can you tell? How can you tell?
It has three sides
That join to form three points.
It's a triangle, it's a triangle.

This is a rectangle, this is a rectangle,
How can you tell? How can you tell?
It has two short sides,
And it has two long sides.
It's a rectangle, it's a rectangle.

Jeanne Petty

I Have A Shape
Sung to: Skip to My Lou

I have a square shape
How about you?
I have a square shape
How about you?
I have a square shape
How about you?
Hold up your square — like I do!

Pass out paper shapes to your children. Give each
child a square, a circle and a triangle. Have everyone hold up the same shape and sing about it

Jean Warren

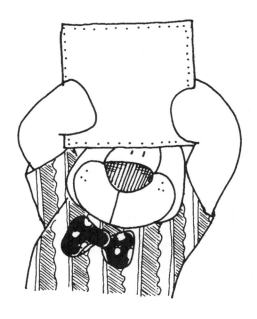

Four Red Apples

Sung to: "This Old Man"

Four red apples on the tree,
Two for you and two for me.
So, shake that tree and watch them fall.
One, two, three, four — that is all.

Additional verses: Four green apples on the tree;
Four brown walnuts on the tree; Four pink blossoms on the tree; etc.

Jean Warren

Climb Aboard the Spaceship

Sung to: "Eensy, Weensy Spider"

Climb aboard the spaceship,
We're going to the moon.
Hurry and get ready,
We're going to blast off soon.
Put on your helmets
And buckle up real tight.
Here comes the countdown,
Let's count with all our might.
10-9-8-7-6-5-4-3-2-1 — Blast Off!

Elizabeth McKinnon

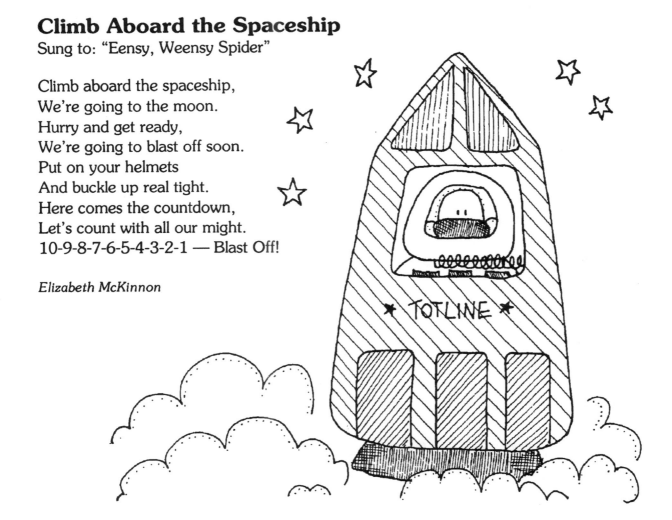

Clock Song
Sung to: "Hickory, Dickory, Dock"

Hickory, dickory, dock,
The time is 1 o'clock.
See the little hand
Point to 1,
Hickory, dickory, dock.

Hickory, dickory, dock,
The time is 2 o'clock.
See the little hand
Point to 2,
Hickory, dickory, dock.

Hickory, dickory, dock,
The time is 3 o'clock.
See the little hand
Point to 3,
Hickory, dickory, dock.

Continue, naming hours up to 12 o'clock. As you
sing, hold up a play clock and move the hands to
the appropriate times.

Elizabeth McKinnon

Count the Days
Sung to: "Twinkle, Twinkle, Little Star"

Come along and count with me,
There are seven days, you see.
Monday, Tuesday, Wednesday too,
Thursday, Friday, just for you.
Saturday, Sunday, that's the end,
Now let's sing it all again!

Judy Hall

Playdough Numerals

When your children are working with playdough, set out numerals that have been cut from cardboard. Encourage the children to roll playdough pieces into snakes. Have them copy the cardboard numerals by forming the snakes into numeral shapes on the table. Or have them arrange the playdough snakes directly on top of the cardboard numerals.

Numeral Rubbings

Cut numeral shapes out of cardboard or sandpaper. Attach the numerals to a tabletop with loops of masking tape rolled sticky sides out. Let your children take turns placing pieces of construction paper on top of the numeral shapes and rubbing over them with crayons.

Opportunities for Learning Math

Stringing Necklaces

Cut plastic straws into short sections. Set them out along with different colored stringing beads. Give each child a long piece of yarn with one end tied to a straw section and the other end taped to form a "needle." Let your children string the straw sections and beads on their yarn pieces as you give directions such as these: "String one straw section; String two beads; String two straw sections; String three beads; String two straw sections; String two beads; String one straw section." When the children have finished, encourage them to count the straw sections and beads on their necklaces. Later, unstring the necklaces and use the pieces again to create new patterns, if desired.

Variation: Instead of giving your children oral directions, provide them with pattern cards to follow as they string their necklaces.

Paper Chains

Choose two colors of construction paper such as red and white. Cut the paper into a number of 1- by 8-inch strips. Let your children work individually or in groups. Have them glue the strips together in different color patterns to create paper chains. For example, suggest that they make alternating loops of red and white paper strips. Or have them follow a pattern such as two red, one white, two red, one white, etc. As the children become familiar with the activity, let them make up color patterns of their own to follow.

Textured Numerals

Give each child a piece of construction paper on which you have drawn a large numeral. Pour glue into shallow containers and set out cotton balls. Have your children dip the cotton balls into the glue and then place the balls along the lines on their papers to create textured numerals.

Variation: Instead of using cotton balls, let your children glue on such things as pasta shapes, popcorn pieces or yarn segments.

Star Numbers

For each child draw a numeral, such as 5, and a horizontal "number line" on a piece of construction paper. Give the papers to your children and set out star stickers. Have the children identify the numeral 5 on their papers. Then let them each count out five star stickers and attach them to their number lines. (Using a number line makes it easier for a child to know where to begin when counting.) Follow the same procedure for other numbers you are working on.

Variation: Set out ink pads and rubber stamps in the shapes of toys or other familiar objects. Give each child a strip of paper with a numeral written at the left-hand end. Let the child choose a rubber stamp and use it to print that number of toys along the length of the strip. Continue with other numerals.

Number Collages

Collect five different kinds of collage materials such as cotton balls, pasta shapes, buttons, toothpicks and tissue paper squares. Place the materials in five separate containers and number the containers from 1 to 5. Hand out pieces of heavy paper or cardboard. Have each child remove the exact number of items from the containers as the numbers indicate. Then let your children glue the items on their papers in designs any way they wish.

Variation: Set out a large piece of paper and write a numeral, such as 3, all over it. Let your children decorate the paper by gluing on small objects (paper clips, toothpicks, circle stickers, etc.) in sets of threes.

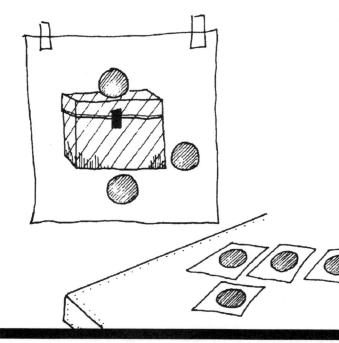

Directions Pictures

To review spatial relationships, occasionally let your children create pictures by following directions. For example, give each child a piece of paper on which you have drawn a picture of a toy box. Also provide each child with several large circle stickers to represent balls. Then give directions such as these: "Place one ball on top of the toy box; Place one ball next to the toy box; Place one ball below the toy box; Place one ball in the middle of the toy box." Encourage the children to compare and discuss their completed pictures.

Housekeeping Activities

Sorting Silverware — Place forks, table knives, serving spoons and teaspoons in a dishpan. After your children wash and dry the silverware, encourage them to sort the pieces into a silverware drawer organizer. Or have them put the knives, forks, serving spoons and teaspoons in separate piles.

Setting a Table — Provide your children with plastic dishes and silverware. Show them a correct place setting that includes a plate, a cup, and a knife, fork and spoon. Then ask the children to follow the pattern you created to make additional place settings on the table.

Sewing Corner — Provide your children with scraps of different textured or colored fabric to match and sort. Also, set out a box containing a variety of buttons. Encourage the children to sort and count the buttons in different ways: by size, by number of holes, by shape, by color, etc.

Clothesline Fun — Set up a clothesline for your children. Provide a basket of different colored fabric squares and some spring-type clothespins. Let the children hang fabric squares on the clothesline and count them. Or make color pattern cards for the children to follow when hanging up the squares. At other times, ask the children to just clip clothespins to the line in sets of twos, threes, etc. Or let them use the clothespins for adding and subtracting activities.

Cooking Area — Provide your children with measuring cups, measuring spoons, bowls and pans to use for pretend cooking activities. Also, keep a set of picture recipe cards on hand. Let the children "read" the recipes and discuss making dishes when they are engaged in imaginary play. Occasionally, create a card for a simple no-cook recipe that the children can actually follow to make a small treat.

Bakery Shop Activities

Making Cookies — Provide your children with different colors of playdough. Have them use cookie cutters (or just their hands) to create cookies of various shapes. When they have finished, encourage them to sort and count their cookie creations.

Making Muffins — Set out a 6-cup muffin tin and a container of playdough. Ask your children to tell you how many balls of dough they will need to fill the muffin tin. Then have them roll out six balls of playdough and count as they place the balls in the muffin tin cups.

Birthday Cake Candles — Let your children pretend that it is someone's birthday. Have them celebrate by making a playdough cake and placing it in a small pan. Give them birthday candles (or plastic straw segments). Encourage the children to count as they place the appropriate number of candles on the cake. Or put the candles on the cake yourself and have the children tell you how old the "birthday person" is.

Bread Sticks — Encourage your children to roll out long playdough snakes. Let them create "bread sticks" by dividing the snakes into pieces. Provide the children with a ruler to use for "measuring" the bread sticks. Have them place the long bread sticks in one container and the short ones in another.

Buying and Selling — When your children want to "sell" bags or boxes of their baked goods, let them attach price stickers that have been numbered from 1 to 5. Provide green construction paper "dollars" (or use play money). When the shoppers come to buy baked goods, have them identify the numerals on the price stickers and pay out that number of dollars. (Use this activity with any kind of play store.)

Fast-Food Restaurant Activities

Sorting Cups — Set out small, medium and large paper cups. Have your children line up the cups from smallest to largest. Then let them sort and count the cups by size into three groups.

Measuring Soft Drinks — Collect a set of different sized paper cups from a fast-food restaurant. Provide water and a set of plastic measuring cups. Let your children experiment with pouring and measuring to discover how much "soft drink" each cup holds.

Making a Hamburger — Give your children paper shapes to represent hamburger ingredients (buns, meat patties, lettuce leaves, tomato slices, cheese slices, etc.). As a cook makes a pretend hamburger, have him or her tell what goes on the bun first, second, and so on.

Counting Fries — Collect several cardboard French fry holders from fast-food restaurants. Cut yellow sponges into french fry shapes. Let your children count the number of fries that will fit in each holder. Or ask the children to fill the holders with specific numbers of fries.

Filling Orders — Let your children take turns requesting and filling pretend food orders (two burgers, one fries and one soft drink, etc.) Provide bags along with props such as paper cups and cardboard food containers. (Or use paper shapes or blocks to represent food items). Encourage the children to count out loud as they fill the orders.

Home Repair Activities

Measuring Tapes — Provide your children with measuring tapes. If possible, include several that roll up automatically. Encourage the children to use the tapes to measure such things as a table's length, a bookshelf's height or a chair's width.

Sequence Cards — Use posterboard squares to make sets of sequence cards that show how to do simple home repairs. For example, draw pictures on the squares illustrating the steps involved in fixing a broken chair leg or changing a burned-out light bulb. Store each set of cards in a separate envelope. Let your children take out the cards and arrange them in the proper order.

Sorting Screws — Set out a muffin tin and a box containing several different kinds of screws. Let a child sort the screws into the cups of the muffin tin. Then have the child count the number of screws in each cup.

Matching Fun — Provide your children with a box of locks and matching keys to play with. Can they find a key to fit each lock? Also provide extra-large nuts and bolts for the children to match and screw together. Or set out two identical sets of paint chips and let the children try matching them by color.

Following Directions — Give directions while your children are doing carpentry projects. For example, say: "Count out four nails. Next, pound one nail into each corner of this piece of wood." Or talk about what you do first, second, and so on, as you assemble a flattened box or open a card table.

Simple Manipulatives

Provide your children with many different kinds of small toys and other objects to use for counting and sorting (make sure the objects are large enough to prevent accidental swallowing.) Some examples would be plastic animals and people, toy cars, dominoes, rocks, poker chips, cardboard shapes and small blocks. Encourage the children to group the objects into different numbered sets or to place them in different numbered containers. Or mix up several kinds of manipulatives and ask the children to sort them into like groups.

Extension: Give your children games that can be used for counting and sorting, such as pegboards with pegs or magnetboards with small magnets.

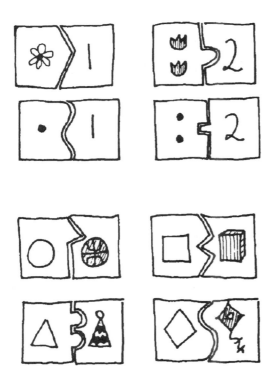

Puzzles

Use a set of several large index cards to make puzzles for your children to play with. For number puzzles, draw a numeral on the right-hand side of each card and a matching number of small pictures (or dots) on the left-hand side. For shape puzzles, draw a geometric shape on the left-hand side of each card and a matching shaped picture on the right-hand side. Cut each card in half to create two interlocking puzzle pieces. (Make sure that each puzzle fits together differently.) To play, mix up all the pieces and let the children take turns putting the puzzles back together.

Blocks

Provide your children with a wide variety of blocks to use for sorting, stacking and counting activities. Also, reinforce skills in such areas as measuring, matching and balancing by encouraging the children to build different kinds of block structures. Or use blocks to help the children understand spatial relationships by giving directions such as these: "Put a block on top of the table; Put a block under the table; Put a block next to the table; Put a block behind the table."

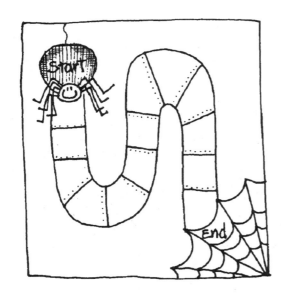

Board Games

Use large pieces of posterboard to make gameboards like the one in the illustration. On each gameboard draw a pathway, divided into spaces, and indicate where to start and finish. Give several children each a marker to place at the beginning of the pathway. Let one child start by rolling a die, counting the dots that come up and moving his or her marker that number of spaces along the pathway. (Or have the child spin a spinner that is labeled with numerals). Then continue in the same manner, letting the players take turns, until everyone has reached the end of the pathway.

Number Lotto

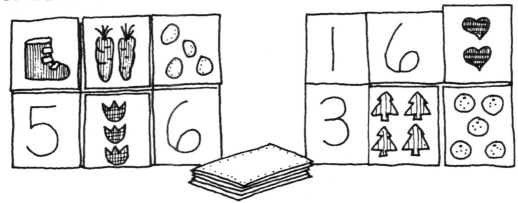

Make several gameboards out of heavy paper. Divide each gameboard into six (or more) sections and label each section with a numeral. For each gameboard make a set of six game cards that are the same size as the numbered sections. Draw matching numbered sets of pictures on the game cards (or attach matching sets of stickers.) To play, give the gameboards to your children and place all the game cards in a deck face down. Let the children take turns drawing cards, counting the numbers of pictures of them and placing them on top of the matching numerals on their gameboards. If a child draws a card that matches one already on his or her gameboard, have the child return the card to the deck and skip that turn. When a player wins by being the first to cover all the sections on his or her gameboard, collect the cards and start the game again.

Card Games

Let your children have fun playing simple card games. For example, invite several children to join you for a game of Pairs. Using an ordinary deck of cards, create a new deck that contains cards numbered from 2 to 6, 8 or 10. Deal four cards to each player. Have the players lay out on the table any pairs of numbered cards they may have. Then have them take turns drawing enough cards from the deck to make five cards in their hands and laying down new pairs of numbered cards. When all the cards have been played, the person with the greatest number of pairs on the table wins.

Number Strips

Cut three 1- by 5-inch strips out of white posterboard. Divide each strip into 1-inch sections. Starting at the left- hand end of each strip, color in each alternate section with a black felt-tip marker. Cut one strip into two shorter strips that are 1 inch and 4 inches long. Cut another strip into one 2-inch strip and one 3-inch strip. Mix up all the strips and give them to a child to play with for a while. Then do activities such as these:

- Ask the child to line up the strips from longest to shortest.

- Point to the 5-inch strip. Ask the child to put together two strips that equal that length. Do the same with the 3-inch strip.

- Put the 3-inch strip and the 1-inch strip together. Ask the child to find one strip that is the same length.

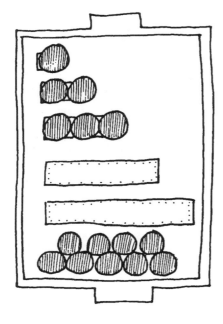

Number Magnetboard

Cut fifteen small circles out of posterboard. Turn the circles into magnets by attaching pieces of magnetic tape to the backs. On a cookie sheet attach five lines of masking tape. Make the first line long enough to hold just one of the magnets, the second line long enough to hold just two of the magnets, and so on. Set out the cookie sheet and the magnets. Let your children take turns placing the magnets on the lines of tape and counting the number that fit on each line.

Cleanup Activities

- Have your children sort their toys by kind before putting them away.

- Mark shelves with shapes of blocks that you have in your room. Have your children match the blocks with the shapes on the shelves to find their proper storage places.

- Ask each of your children to pick up a specific number of papers from the art area or toys from the play area.

- When cleanup time begins, set a kitchen timer to go off after 3 or 4 minutes. Have your children try to straighten up the room and be seated in a circle before the timer rings.

Lining-Up Fun

Help develop sorting and classification skills by having your children line up in different ways. Following are a few examples.

- From youngest to oldest.

- From shortest to tallest.

- By number of buttons on clothes (all children with no buttons, all children with one button, and so on).

- By color of shoes (all children with red shoes, all children with white shoes, etc.).

- By birthdates (all children with birthdays in January, all children with birthdays in February, and so on).

- By kind of clothing (all children wearing mittens, all children not wearing mittens; all children wearing belts, all children not wearing belts; etc.).

Waiting Games

Finger Counting Game — Ask your children to put one of their hands behind their backs and stick out one to five fingers, while you do the same. Have the children bring out their hands with their fingers extended. After they have done so, reveal your hand and count the number of fingers you are holding up. Any children who are holding up that same number of fingers are the winners.

I'm Thinking of a Number — Start by saying something like this: "I'm thinking of a number that comes after three and before five. What is it?" When your children have correctly named the number, continue with clues for other numbers, such as these: "I'm thinking of a number that tells what day it is today; I'm thinking of a number that tells how old Jennifer is; I'm thinking of a number that tells how many shoes I am wearing."

Tracing Numerals With Fingers — Try this game whenever your children have to wait in line for a long time. Have them form into pairs (join the line yourself if you have an uneven number of children). Then let the children take turns drawing numerals on each other's backs for their partners to guess.

Numbers Up — Have your children stand in a line. Call out a game number between 1 and 10. Explain that the point of the game is to avoid saying that number. Starting at the beginning of the line, have the children count in numerical order. Each child may name one or two consecutive numbers. When a child has to say the game number, have him or her sit down. Then let the next child in line start counting again at number one. Continue the game until only one child is left standing. Let that child name a new game number, if desired.

Notes:

Notes:

Totline® Books

For parents, teachers, and others who work with young children

TEACHING THEMES

THEME-A-SAURUS®

Classroom-tested, around-the-curriculum activities organized into imaginative units. Great for implementing a child-directed program.

Theme-A-Saurus

Theme-A-Saurus II

Toddler Theme-A-Saurus

Alphabet Theme-A-Saurus

Nursery Rhyme Theme-A-Saurus

Storytime Theme-A-Saurus

BUSY BEES SERIES

Designed for two's and three's—these seasonal books help young children discover the world through their senses. Activity and learning ideas include simple songs, rhymes, snack ideas, movement activities, and art and science projects.

Busy Bees—SPRING

Busy Bees—SUMMER

Busy Bees—FALL

Busy Bees—WINTER

PLAY & LEARN SERIES

This creative, hands-on series explores the versatile play-and-learn opportunities of familiar objects.

Play & Learn with Stickers

Play & Learn with Paper Shapes and Borders

Play & Learn with Magnets

Play & Learn with Rubber Stamps

Play & Learn with Photos

GREAT BIG THEMES

Giant units that explore a specific theme through art, language, learning games, science, movement activities, music, and snack ideas. Includes reproducible theme alphabet cards and patterns.

Space

Farm

Zoo

Circus

CELEBRATIONS SERIES

Easy, practical ideas for celebrating holidays and special days around the world. Plus ideas for making ordinary days special.

Small World Celebrations

Special Day Celebrations

Great Big Holiday Celebrations

EXPLORING SERIES

Encourage exploration with hands-on activities that emphasize all the curriculum areas.

Exploring Sand and the Desert

Exploring Water and the Ocean

Exploring Wood and the Forest

NUTRITION

SNACKS SERIES

This series provides easy and educational recipes for healthful, delicious eating and additional opportunities for learning.

Super Snacks

Healthy Snacks

Teaching Snacks

Multicultural Snacks

LANGUAGE

CUT & TELL CUTOUTS

Each cutout folder includes a delightful tale, color figures for turning into manipulatives, and reproducible activity pages.

COLOR RHYMES *Rhymes and activities to teach color concepts.*

Cobbler, Cobbler

Hickety, Pickety

Mary, Mary, Quite Contrary

The Mulberry Bush

The Muffin Man

The Three Little Kittens

NUMBER RHYMES *Emphasize numbers and counting.*

Hickory, Dickory Dock

Humpty Dumpty

1, 2, Buckle My Shoe

Old Mother Hubbard

Rabbit, Rabbit, Carrot Eater

Twinkle, Twinkle, Little Star

NURSERY TALES *Enhance language development with these classic favorites.*

The Gingerbread Kid

Henny Penny

The Three Bears

The Three Billy Goats Gruff

Little Red Riding Hood

The Three Little Pigs

The Big, Big Carrot

The Country Mouse and the City Mouse

The Elves & the Shoemaker

The Hare and the Tortoise

The Little Red Hen

Stone Soup

TAKE-HOME RHYME BOOKS SERIES

Make prereading books for young children with these reproducible stories. Great confidence builders!

Alphabet & Number Rhymes

Color, Shape & Season Rhymes

Object Rhymes

Animal Rhymes

MUSIC

PIGGYBACK® SONGS

New songs sung to the tunes of childhood favorites. No music to read! Easy for adults and children to learn. Chorded for guitar or autoharp.

Piggyback Songs

More Piggyback Songs

Piggyback Songs for Infants & Toddlers

Piggyback Songs in Praise of God

Piggyback Songs in Praise of Jesus

Holiday Piggyback Songs

Animal Piggyback Songs

Piggyback Songs for School

Piggyback Songs to Sign

Spanish Piggyback Songs

More Piggyback Songs for School

Totline Books are available at local parent and teacher stores

TEACHING RESOURCES

BEAR HUGS® SERIES

Think you can't make it through another day? Give yourself a Bear Hug! This unique series focuses on positive behavior in young children and how to encourage it on a group and individual level.

Meals and Snacks
Cleanup
Nap Time
Remembering the Rules
Staying in Line
Circle Time
Transition Times
Time Out
Saying Goodbye

Saving the Earth
Getting Along
Fostering Self-Esteem
Being Afraid
Being Responsible
Being Healthy
Welcoming Children
Accepting Change
Respecting Others

1001 SERIES

These super reference books are filled with just the right tip, prop, or poem for your projects.

1001 Teaching Props
1001 Teaching Tips
1001 Rhymes & Fingerplays

THE BEST OF TOTLINE®

A collection of the best ideas from more than a decade's worth of Totline Newsletters. Month-by-month resource guides include instant, hands-on ideas for around-the-curriculum activities. 400 pages

LEARNING & CARING ABOUT SERIES

Developmentally appropriate activities to help children explore, understand, and appreciate the world around them. Includes reproducible parent flyers.

Our World
Our Selves
Our Town

MIX AND MATCH PATTERNS

Simple patterns, each printed in four sizes.

Animal Patterns
Everyday Patterns
Nature Patterns
Holiday Patterns

1•2•3 SERIES

Open-ended, age-appropriate, cooperative, and no-lose experiences for working with preschool children.

1•2•3 Art
1•2•3 Games
1•2•3 Colors
1•2•3 Puppets
1•2•3 Reading & Writing
1•2•3 Rhymes, Stories & Songs
1•2•3 Math
1•2•3 Science
1•2•3 Shapes

101 TIPS FOR DIRECTORS

Great ideas for managing a preschool or daycare! These hassle-free, handy hints help directors juggle the many hats they wear.

Staff and Parent Self-Esteem
Parent Communication
Health and Safety
Marketing Your Center
Resources for You and Your Center
Child Development Training

FOUR SEASONS SERIES

Each book in this delightful series provides fun, hands-on activity ideas for each season of the year.

Four Seasons–Movement
Four Seasons–Science

PARENTING RESOURCES

A YEAR OF FUN

These age-specific books provide information about how young children are growing and changing and what parents can do to lay a strong foundation for later learning. Calendarlike pages, designed to be displayed, offer developmentally appropriate activity suggestions for each month—plus practical parenting advice!

Just for Babies
Just for One's
Just for Two's
Just for Three's
Just for Four's
Just for Five's

TEACHING HOUSE SERIES

This new series helps parents become aware of the everyday opportunities for teaching their children. The tools for learning are all around the house and everywhere you go. Easy-to-follow directions for using ordinary materials combine family fun with learning.

Teaching House
Teaching Town
Teaching Trips

CHILDREN'S STORIES

Totline's children's stories are called Teaching Tales because they are two books in one—a storybook and an activity book with fun ideas to expand upon the themes of the story. Perfect for a variety of ages. Each book is written by Jean Warren.

Kids Celebrate the Alphabet
Ellie the Evergreen
The Wishing Fish
The Bear and the Mountain

HUFF AND PUFF® AROUND THE YEAR SERIES

Huff and Puff are two endearing, childlike clouds that will take your children on a new learning adventure each month.

Huff and Puff's Snowy Day
Huff and Puff on Groundhog Day
Huff and Puff's Hat Relay
Huff and Puff's April Showers
Huff and Puff's Hawaiian Rainbow
Huff and Puff Go to Camp
Huff and Puff on Fourth of July
Huff and Puff Around the World
Huff and Puff Go to School
Huff and Puff on Halloween
Huff and Puff on Thanksgiving
Huff and Puff's Foggy Christmas

Totline Books are available at local parent and teacher stores

Active preschool learning— Ideas that work!

From Totline® Publications

Totline® MAGAZINE

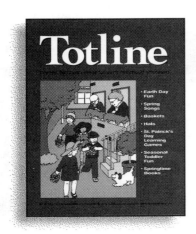

Now in full color!

Challenge and engage young children with the fresh ideas for active learning in *Totline Magazine*. Developed with busy, early-childhood professionals and parents in mind, these activities need minimal preparation for successful learning fun. Each bimonthly issue is perfect for working with children ages two to six and includes • seasonal learning themes • stories, songs, and rhymes • open-ended art projects and science explorations • reproducible parent pages • ready-made teaching materials • and activites just for toddlers. *Totline Magazine* is the perfect resource for a project-based curriculum in a preschool or at home.

Reproducible! Super Snack News

This delicious newsletter is meant to be shared!

Make up to 200 copies per issue with each subscription, then use the copies as informative handouts! *Super Snack News* is a monthly, four-page newsletter featuring healthy recipes that parents and preschoolers will love, learning activities perfect for the home environment, plus nutrition tips. Also provided are category guidelines for the Child and Adult Care Food Program (CACFP). Sharing *Super Snack News* is a wonderful way to help promote parent involvement in quality childcare.